History, Philosophy and Theory of the Life Sciences

Volume 39

History, Philosophy and Theory of the Life Sciences is a space for dialogue between life scientists, philosophers and historians – welcoming both essays about the principles and domains of cutting-edge research in the life sciences, novel ways of tackling philosophical issues raised by the life sciences, as well as original research about the history of methods, ideas and tools, which constitute the genealogy of our current ways of understanding living phenomena.

The series is interested in receiving book proposals that • are aimed at academic audience of graduate level and up • combine historical and/or philosophical and/or theoretical studies with work from disciplines within the life sciences broadly conceived, including (but not limited to) the following areas: • Anatomy & Physiology • Behavioral Biology • Biochemistry • Bioscience and Society • Cell Biology • Conservation Biology • Developmental Biology • Ecology • Evolution & Diversity of Life • Genetics, Genomics & Disease • Genetics & Molecular Biology • Immunology & Medicine • Microbiology • Neuroscience • Plant Science • Psychiatry & Psychology • Structural Biology • Systems Biology • Systematic Biology, Phylogeny Reconstruction & Classification • Virology The series editors aim to make a first decision within 1 month of submission. In case of a positive first decision the work will be provisionally contracted: the final decision about publication will depend upon the result of the anonymous peer review of the complete manuscript. The series editors aim to have the work peer-reviewed within 3 months after submission of the complete manuscript. The series editors discourage the submission of manuscripts that contain reprints of previously published material and of manuscripts that are below 150 printed pages (75,000 words). For inquiries and submission of proposals prospective authors can contact one of the editors: Charles T. Wolfe: ctwolfe1@gmail.com Philippe Huneman: huneman@wanadoo.fr Thomas A.C. Reydon: reydon@ww.uni-hannover.de

Victoria Shmidt · Simone De Angelis
Editors

Blumenbach's Racial Classification

Deconstructing the Timeless Call for Human Differentiation

FWF Österreichischer Wissenschaftsfonds

Springer

Editors
Victoria Shmidt
Centre for the History of Science
University of Graz
Graz, Austria

Simone De Angelis
Centre for the History of Science
University of Graz
Graz, Austria

ISSN 2211-1948 ISSN 2211-1956 (electronic)
History, Philosophy and Theory of the Life Sciences
ISBN 978-3-032-15949-6 ISBN 978-3-032-15950-2 (eBook)
https://doi.org/10.1007/978-3-032-15950-2

Published with the support of the Austrian Science Fund (FWF): 10.55776/PUB1279; Research results from Austrian Science Fund (FWF); 10.55776/V1020

This Springer imprint is published by the registered company Springer Nature Switzerland AG
The registered company address is: Gewerbestrasse 11, 6330 Cham, Switzerland

I found this collection uniformly excellent in this respect. It brings together recent work on eighteenth and nineteenth century discussions of the concept of race and the work on Blumenbach's importance for these discussions. In this it has utilized the resources of the Blumenbach archival work and also engaged contemporary discussions of methodology and historiography in relation to the historical assessment of Blumenbach's longer influence up to the present. The work is impressive in this respect and I learned a good deal from this aspect of the work.

—Phillip R. Sloan

In conclusion, the volume is a collection of case studies that has not only analytical but also ethical value: on the one hand, it provides a better understanding of how certain scientific theories are constructed and how their dissemination in different societies actually takes place. On the other hand, it lets reflect critically on the mechanisms of power and the responsibilities of the man of science: Indeed, the major merit of the volume consists in historical and theoretical explanations of the role played both by academic affiliations and political processes (colonialism, the birth and strengthening of national states) in the transformation of cognitive constructions and practices of great social and political relevance such as racial classifications.

—Mario Marino

Preface

This volume explores how the elements of Johann Friedrich Blumenbach's epistemology have been variously assembled in different ways and in different historical-cultural and political-social contexts. It looks at the ways in which this epistemology has been enriched with new elements that have ensured its reproduction. Blumenbach is one of the first thinkers whose legacy opened up space for a mutually controversial historicization of the scientist's impact on scientific racism and the fight against it. The contributors revise approaches to the historical legacy of Enlightenment scholars such as Blumenbach in terms of nuances to our understanding of intellectual responsibility for the reproduction of race-informed thought, the legitimization of racial discrimination, and the lack of critical response to both. *Blumenbach's Racial Classification: Deconstructing the Timeless Call for Human Differentiation* transcends the simplistic boundaries between "right" and "left" science and revises the internally contradictory impact of racialist/racial thinking on knowledge about humanity.

Graz, Austria

Victoria Shmidt
Simone De Angelis

Acknowledgements

As those who have undertaken the task of revising Blumenbach's legacy, we are infinitely grateful to the authors and implementers of *Johann Friedrich Blumenbach—Online*, the project that has digitalized his archives, especially his correspondence. This project has provided us with a unique source of new positive provocations and challenges. The editors thank all the contributors for their commitment to the basic idea underlying this collection.

Cosmin Koszor-Codrea, Kseniya Brailovskaya, Malin Sonja Wilckens, and Victoria Shmidt had the chance to present their papers to the International Philosophy of Biology Circle in Fall 2024, which provided a unique opportunity to receive feedback and to deepen their arguments. Reviews by Catherine Baker, Mario Marino, and Phillip Sloan navigated editors and contributors to find the balance between making the book more friendly to readers and introducing new ways of thinking about race in science. Sahotra Sarkar's comments helped deepen our argument in the introduction. The editors thank Charles Wolfe for inspiring them to undertake this project and Christopher Wilby for his continued support at every stage of its publication. Ivana Prazic, Khairunnisa A Shukor, Sara Shakilla Mohd Salim, and Shakila Che Dahalan are grateful to Asif Mohi Ud Din for his comments on an early draft of their text.

The publication of the book is a research results from: The Austrian Science Fund (FWF): project "A Critical History of Genetics and its Implications in CEE countries" https://doi.org/10.55776/V1020

Acknowledgments

Contents

Editors and Contributors

About the Editors

Victoria Shmidt is senior postdoc researcher at the Centre for History of Science at the University of Graz, Austria and principle investigator of the Project A Critical History of Genetics and its Implication in Central Eastern Europe supported by FWF. Her recent publications focus on race in science and its operation in Central Europe: "Historicizing Roma in Central Europe Between Critical Whiteness and Epistemic Injustice" (Routledge, 2021); "A Critical History of Health Films in Central and Eastern Europe and Beyond" (Routledge, 2023); special issue of *Romani Studies* "Epistemolhtogies in Romani studies: Moving beyond othering otherness" (2024).

Simone De Angelis is Professor of History of Science and Director of the Centre for History of Science at the University of Graz, Austria. His main research areas are Renaissance science and early modern history and philosophy of science and medicine, Enlightenment studies, history of life sciences, and anthropology. He has been Fellow of the Swiss National Science Foundation, the Herzog August Bibliothek in Wolfenbüttel, Visiting Researcher at the Wellcome Trust Centre for the History of Medicine at University College London and the Università degli studi di Padova. He is a Visiting Professor at École des Hautes Études en Sciences Sociales (EHESS) and the Centre Alexandre Koyré (CAK) in Paris. His books include Von Newton zu Haller (Niemeyer 2003), Scholarly Knowledge. Textbooks in early modern Europe (Droz 2008), Anthropologien (De Gruyter 2019, 2nd edition).

Contributors

Simone De Angelis Centre for the History of Science, University of Graz, Graz, Austria

Kseniya Brailovskaya Centre for the History of Science, University of Graz, Graz, Austria

Shakila Che Dahalan The Faculty of Human Sciences, Sultan Idris Educational University (UPSI), Tanjung Malim, Malaysia

Cosmin Koszor-Codrea New Europe College, Bucharest, Romania

Ivana Pražić The Faculty of Human Sciences, Sultan Idris Educational University (UPSI), Tanjung Malim, Malaysia

Sara Shakilla Mohd Salim The Faculty of Human Sciences, Sultan Idris Educational University (UPSI), Tanjung Malim, Malaysia

Victoria Shmidt Centre for the History of Science, University of Graz, Graz, Austria

Khairunnisa A Shukor The Faculty of Human Sciences, Sultan Idris Educational University (UPSI), Tanjung Malim, Malaysia

Malin S. Wilckens Leibniz Institute of European History, Mainz, Germany

List of Figures

Chapter 1
Introduction: The Recontextualization of Blumenbach's Legacy in Uncertain Times

Victoria Shmidt and Simone De Angelis

Abstract In this introductory chapter, the recent and current historicization of Blumenbach's legacy is discussed as the interplay between reductive contextualism and critical responses to it. Three interrelated contexts shape the historicization of Blumenbach as fluctuating between attempts to fix his legacy and to critically examine the driving forces behind the ongoing reproduction of the Blumenbachian approach: (1) race in science and the impact of science on producing both racist and anti-racist arguments; (2) the transfer between premodernity (the eighteenth-century or theological approach to humanity) and modernity (the nineteenth-century or secular/scientific vision of science); and (3) the public or political engagement of scholars in (de)legitimizing particular politics against human dignity. Relying on Hans Blumenberg's approach to exploring the complex nexus of pre-modernity—modernity and Sylvia Winter's application of Blumenberg's vision to the issue of whiteness, we develop a framework for the immanent critique of race in science. Following the longue durée of Blumenbach's approach to human variation as numerous adaptations provides options for revising the entangled history of race in science.

Keywords Reductive contextualism · Contextualization · Premodernity · Modernity · Adaptation · Critique of racism · Historicization · Race in science · Exemplary narrative · Political engagement of scholars

V. Shmidt (✉) · S. De Angelis
Centre for the History of Science, University of Graz, Graz, Austria
e-mail: victoria.shmidt@uni-graz.at

S. De Angelis
e-mail: simone.de-angelis@uni-graz.at

V. Shmidt and S. De Angelis (eds.), *Blumenbach's Racial Classification*, History, Philosophy and Theory of the Life Sciences 39,
https://doi.org/10.1007/978-3-032-15950-2_1

1.1 Historicizing Blumenbach's Legacy Through the Challenges of Reductive Contextualism

The ongoing revision of Johann Friedrich Blumenbach's legacy reflects one of the core principles in critical historicization, namely, that the changing present redefines the past. The logic underlying the reproduction of interest in Blumenbach and his work is primarily determined by the debates about the history of the biological arguments for and against racism (Rupke, 2018, 234). The infinite horizon of such historicization is, in turn, determined by the constantly expanding body of knowledge about the origins of "man" and doubts about its reliability. Blumenbach's sociological biography is constantly being updated with new twists—thanks to the fact that both Blumenbach's texts and the well-documented life of the researcher remain valuable sources for critical reflection, as well as political speculation. Further, recent actions surrounding Blumenbach's legacy, such as displaying his bust in the public environment at the University of Göttingen, implementing the project *Sensible Provenienzen* (2020–2023) aimed at investigating the origins of Blumenbach's skull collection, and returning its remnants (Mareike, 2023) call for connecting the authorship of racial thought and its circulation in various areas of public policy.

Methodologically, revising Blumenbach's legacy challenges the recognition of the history and philosophy of science through the lenses of modern and post-modern approaches. If "Culture and history define use, but cannot themselves be defined in terms of usefulness," (Ankersmit, 1989, 139) does history define justice? Can historical justice regarding Blumenbach's legacy fall into line with transitional justice for those against whom race-informed practices were applied and reconsider contemporary approaches to studying human variation? Is revising Blumenbach's legacy meaningful (and useful) for overcoming multiple forms of epistemic injustice toward those objectified by race science? These questions raise the task of contextualization—particularly complicated in the case of Blumenbach's legacy. While race in science is one of the streams in history of science connected with the mission to produce significant narratives, the question asked by Hayden White (1998, 22)—"What are historical representations representations of?"—can be paraphrased as "What are historical contexts contexts of?".

In consonance with Dominick LaCapra (2013, 17), we problematize the legacy of Blumenbach in terms of reductive contextualism, which relegates the work and play of the text to the margins of one or another chronology, serving one or another historical representation. The collision of "double sensitivity," central to contextualism, concerns, sensitivity to the standards of knowledge explored, on the one hand, and sensitivity to one's own process of knowledge production on the other (Richard, 2004). This collision generates multiple challenges for exploring the past of science. Historicizing the legacy of Blumenbach as a "parental figure" of anthropology represents the widest possible range of such challenges.

The interpretation of Blumenbach's impact is often aligned with recognizing his embeddedness in the inception of anthropology, which started "with the scandal of domination and exploitation of one part of mankind by another" (Fabian, 2014, x). The exemplary narratives (Rüsen, 2005; Rüsen, 1982) in which Blumenbach is situated as a typical representative of his era, who, due to circumstances, still acts as a figure whose significance should be questioned, are accompanied by the negation of such views as simplified and historically inaccurate.

The copious contextual reductionism of Blumenbach's legacy constitutes a "permanent parabasis," (La Capra 14) which can be explained only partially through the overproduction of historical literature in a time of post-modern diversification of the past (Ankersmit ibid). Attacking contextual reductionism chiefly targets one or another set of contextual forces,and this selective struggle for critical historicization, against the overflow of reductionist contextualization, should be seen as a source of additional challenges.

Three interrelated contexts shape the historicization of Blumenbach as fluctuating between the attempts to fix his legacy and to critically examine the driving forces behind the ongoing reproduction of the Blumenbachian approach: (1) race in science and the impact of science on producing both racist and anti-racist arguments; (2) the transfer between premodernity (the eighteenth-century or "theological" approach to humanity) and modernity (the nineteenth-century or secular/scientific vision of science); and (3) the public or political engagement of scholars in (de)legitimizing particular politics against human dignity.

Several generations of attempts to explain ***the longue durée of the racially informed production of knowledge*** in general, and Blumenbach's conceptualization of race in particular, attest not only to the intractability of "race science" but also to the complicated development of its critical acceptance. Many recent publications have been targeted at the ultimate negation of racially informed knowledge call for delegitimizing race and racial identity entirely (Hill, 2021). However, neutral or even positive evaluations of "race science," and especially Blumenbach's classification scheme, can be seen as engines for the ongoing application of racial classification in favor of progress in medicine (Rambachan, 2018) or physical anthropology (Bhopal, 2007). Moreover, the broader question of archiving "race science," which has been debated during the recent digitalization of Blumenbach's collection, points to multiple contradictions in interpreting the impact of racial classification on science (Marino, 2022).

Practically, both extremes as regards the historical role of Blumenbach in scientific racism battle with one another through very similar sets of story-telling devices, namely, the indirect use of myth, or what Northrop Frye (1961, 603) defined as "displacement," due to the pressure of the intention to produce "credible, logically motivated and morally acceptable" narratives. Encoding through cuturally proven categories (White, 1998) such as "racism," "scientific fact," or even "progress" not only fulfills the mission of making challenging events more familiar but also decreases anxiety in the face of moral and epistemological relativism regarding race in science. Thomas Junker's attack (1998) on Stephen Jay Gould (2014), who, according to Junker, directly blamed Blumenbach for equipping the Nazis with the racial anthropology, is exemplary.

Junker builds his critique by accusing Gould of engaging in a historically inaccurate, transforming visualization of Blumenbach's racial classification from primary linear structure to a hierarchical figure. Junker designs his argument in terms of the triumph of imagination over mind (Frye, 1961, 596) in Gould's argument, which he explains through Gould's presentism. Obviously, Junker overlooks the close relation between Gould's argument in favor of accepting Blumenbach as one of the pioneers of race science and his approach to historicizing the theorization of the evolutionary process. Gould (1977), who consistently deconstructed any linear narrative of scientific progress in understanding evolution, ranked Blumenbach among those scholars who, by following the canon, only perpetuated a path with limited opportunities for reasonable skepticism and self-criticism. For Gould (1977), who recognized the heuristic role of preformation and denied recapitulation theory, Blumenbach remained one of those who connected pre-racial classifications, such as Linnaeus' typology, and Darwin's extrapolation of his explanation of animal extinction, to humans. Gould's visualization of Blumenbach's interpretation of human diversity thus resonates with the very audible motif of recapitulation in the visualization of different races in Blumenbach's *Beiträge zur Naturgeschichte* (see more in Shmidt's chapter in this volume). Further, in his own interpretation of Blumenbach, Junker (2018) fully applies the contemporization of the historical source (Ankersmit, 1989, 146) in favor of proving Blumenbach's dedication to the idea of human unity. These extremes in approaches to the racially informed production of knowledge reverberate with the growing uncertainty about the independence of antiracist rhetoric as capable of producing knowledge free from the risk of being misused by racially minded experts (Táíwò, 2022).

The historical responsibility for producing race-informed arguments and the reexamination of the Enlightenment as the cradle of modern racism brings Blumenbach's legacy to ***the context of continuities and ruptures between premodernity and modernity.*** Since the publication of the punchy text, "The racial contract," by Charles W. Mills in 1997, delegitimizing Western philosophy and its racist underpinnings has led to the recognition of "race science" as a radical agent, even as one of the main "authors" of Western thought. Mill's historicization combines recognition of the original role of the racial contract in shaping Western civilization since antiquity with a division of the racial contract's operation into two periods: an overt one in times of colonialism and its long-term echoes after that. This temporal duplicity led him to focus on the role of scientific ideas in tacitly subordinating the racial contract even after the era of overt legal enforcement of racial discrimination had ended. Mills suggests a rejection of the division into "bad" scientists, to whom the role of scapegoats for racism is attributed (e.g., Gobineau) and "good" scientists, seen as apologists of anti-racism (Mills, 1997, 117). This rejection implies that Mills' view on Western science is inherently racist, which in combination with accepting its epistemic hegemony, relegates the interrelation between science and other agents of knowledge production—including religion—to the margins of critical historicization.

Historically informed reflections on adapting racial thinking for various purposes by the most prominent philosophers reveal the complicated relationship between science and religion as a driving force underlying the meteoric rise of racial science since the last quarter of the eighteenth century (Keel, 2018). The view on "race science" as a long-term agency of human thinking brings into question the many intentions to eliminate the concept of "race" from social constructions of humanity because the concept is biologically non-existent: "Because race in the biological sense has already been disproved, a belief in race is tantamount to a belief in elves, fairies, and winged horses." (Hill, 2017) In this argumentation, it is easy to recognize the ultimate ignorance of racial thinking as a power structure embedded in many, if not all, realms of social life. Furthermore, this ignorance is also not reflected upon, because it corresponds to the modern, secularized notion of science, especially natural science, as the highest level of professional competence. These two main lines of deconstruction regarding race in science, the complete stigmatization of the social sciences represented in Mill's writings, and the primacy of the natural sciences in developing anti-racist arguments, further complicate the task of historicizing race in science.

The contemporary historicization of Blumenbach brings him into the grip of mutually conflicting approaches to historicizing modernity (Schmidt, 2014). The view on Blumenbach as the father of modern racial anthropology, disseminated since the 1970s, is questioned by those who reveal the complicated relationship between science and religion as a driving force underlying the meteoric rise of racial science since the last quarter of the eighteenth century (Keel, 2018). Being one of the agents of transfer from premodern to modern approaches to the operation of science, Blumenbach is often historicized through placing him into multiple classifications and typologies that aim at exploring the interrelation between teleological and scientific visions of humanity (Niekerk, 2014). While typologies connect mutually contesting approaches with the clusters of names or even geopolitical localities (Zammito, 2018), the positions of particular scholars are predominantly defined through mutual negation and debated in terms of the primary authorship of one or another idea. For instance, historicizing the legacy of Georges Cuvier through putting him into opposition to a wide range of his companions leads to the obviously weak and predictable conclusion regarding Cuvier's eclecticism on the mechanisms of evolution (Outram, 1986).

One of the examples of reducing Blumenbach's legacy to labeling him as a member of particular cluster is recognition of his belonging to "vital materialism," an umbrella concept introduced by Timothy Lenoir for stressing the emancipation of Blumenbach's cohort from its theological background. Indicating Blumenbach as comparable to Johann Christian Reil,[1] Lenoir describes their stance regarding the forces behind life as "not to be conceived as some independent entity but rather as an emergent property dependent upon the specific order and arrangement of the components." (Lenoir, 1989).

By interpreting Blumenbach's legacy through signifiers such as materialism, Lenoir intends to deconstruct "the persistence in adapting Blumenbach's texts "[as] often explained in terms of following the continued reproduction of a teleological framework." (ibid) Lenoir attempts to attribute the dialectical opposition of part and

[1] Johann Christian Reil was a German doctor, a younger contemporary of Blumenbach's, who introduced not only the term "psychiatry", but also adapted the concept of *Bildungstrieb* for the needs of exploring mental health and disease.

whole that interprets human variation as mostly modeling, rather than describing actual state of art to the *Bildungstrieb,* an explanatory scheme introduced by Blumenbach (ibid 21–24). For Lenoir, the transitory character of the interrelation among races, decisive for Blumenbach's temporality in human variation, is lost in favor of coming down on Blumenbach's idea of degeneration as a response to physical causes.

In terms of Hyden White, Lenoir applies the concept of "culturally sanctioned meanings" (White ibid 22) such as "materialism," which assist him in "proving" his vision of natural science and its history through the rupture between premodern and modern times. Relative to this, massive critical response to Lenoir's interpretation has focused on his representation of Blumenbach's legacy, which could be defined vis a vis White as "exclusively provisional and contingent historical."(ibid 16) This negation is embedded in a trend away from binary oppositions aimed at providing a historically simplified vision of the relationship between premodernity and modernity. Charles Wolf revises the boundaries between vitalism and mechanism through the lens of teleomechanism as a form of historical proof that epistemologies have their dialectical development (Wolfe, 2014) and should be explored through such peculiar forms of symbiosis.

In his interpretation of Lenoir's approach as improperly simplified, Philippe Huneman notes a retrospective reconstruction of science explored by Lenoir as subsequent history, the two original instigators of a research program and their divergence (Huneman, 2007). This vision, according to Huneman, can neither embrace such practices as medicine nor what could be defined now as health propaganda. It represents rather a debate on races, the central domain, according to Huneman, for interpreting the complex relationship between pre-modernity and modernity in the epistemologies of Blumenbach and his contemporaries (ibid). Understanding the relationship between mechanistic explanation and vitalist organismatic approaches through the modern transformation of histories of humanity into an instrument of political legitimation (Woodard, 2020) resonates with the trend to follow the relationship between different periods of natural science as a dialogue.

The critical revision of options for applying the concept of vital materialism to exploring the transfer from teleological logics to scientific programs of examining nature culminates in the detailed investigation by Phillip Sloan, who follows the role of Émilie du Châtelet in the inception of vitalism (Sloan, 2019). By placing the definition of "vital" in quotation marks, Sloan not only emphasizes the conventionality of the irreconcilable opposition between the vital and the mechanistic, but also creates the necessary tension between irony and simplified historicization. Sloan deconstructs often taken-for-granted divisions such as mechanisms vs. vitalism, idealism vs. materialism, or preformation vs. epigenesis through following the multiphase transition from a previous, metaphysically grounded teleological theory of life to a materialist thesis about the constitution of living beings through their construction from a special kind of vital matter (ibid 19). The hybrid character of Buffon's views investigated by Sloan questions the "standpoint" trope disseminated in the history of science. Instead of such a trope, Sloan follows how Buffon accommodated the "dynamic relations with the world's multiplicity," (Watson, 2011, 66) including human variation.

A vision of the Darwinian revolution as a rolling process, in which the end of the eighteenth century and the first third of the nineteenth century represent formative years (Herbert, 2005), opens the floor for revising continuities and ruptures of progress in developing the idea that species were not created separately from each other (Depew & Bruce, 2011; Depew, 2017). While a "conversation with Romantic science" (Lambier, 2016) has recently obtained a solid place in historicizing biology in the late nineteenth century, the question of why the next generations of thinkers engaged in dialogue with Romanticism and its representatives calls for extending, or even transforming, the timeline of such a relational approach. With this extension, not only the evolutionary turn (ibid) but a wider range of mutual connections can be recognized in favor of problematizing the statement that "nineteenth-century racism cleared all these contradictions [between mechanism and organicism] by unilaterally developing certain aspects of his doctrine." (Marino, 2022).

Reducing Blumenbach's legacy to the task of demonstrating its relation to racism aggravates the risk of moral relativism. Looking at Blumenbach by using his biography as a source of signifiers for differentiating modernity from premodernity increases epistemological relativism. The interplay of moral and epistemological reductionisms practically blocks the options for recognizing his political engagement: Blumenbach's participation in ***establishing the modern order of the operation of science***, which Hans Blumenberg has explored in terms of legitimacy, the capacity to ground scientific and moral progress in rational insight alone (Ingram, 1990). Apparently, Blumenbach was concerned not only about the debates over slavery, but also with many other issues that divided society at the time and that called for expert opinions. Through Blumenbach's correspondence, it is possible to follow his involvement in multiple public activities, from demonstrating the efficiency of the first attempts to vaccinate children against smallpox (Blumenbach, 1777/2006, 122) to attacking slavery. This part of his life, however, remains largely relegated to the margins, especially in contrast to the debates regarding the ideas that he promoted or attacked. In the narratives aimed at presenting him exceptionally as a creator of modern scientific racism, Blumenbach is oftentimes seen as a "cabinet" scholar whose scholarship stemmed from the colonial practices of forced purchasing. Repeatedly, Blumenbach is pictured as primarily a thinker who adapted and mixed the best ideas of his time, and who remained in the shadow of Buffon and Kant (Goy, 2017; Eberl, 2019). These clichés displace the very important task of understanding how Blumenbach connected the different "floors" of the production and dissemination of a racially informed vision of human history.

Counter-narratives often emphasize Blumenbach's friendships with leaders of the abolitionist movement and his devotion to promoting equality between whites and people of color (Mazzolini, 2018). The examples of a balanced view on Blumenbach's sociological biography primarily address his complex in-betweenness regarding evolutionary thought, which require further effort to relate to Blumenbach's political engagement. One of the exceptional examples of moving beyond the established frames of narration is following the interest in non-Western occult practices such as shamanism, one of the many lines of convolution between Blumenbach and Goethe (Flaherty, 1989). Such narratives introduce serious doubts regarding the straight nexus of pre-modern and modern affiliation of Blumenbach. How can we engage the task to revise Blumenbach's legacy and avoid the risk of collapsing the philosophy of history and historiography because of inevitable reductionism? (Kansteiner, 1993).

1.2 Temporalizing Blumenbach's Legacy: Toward an Immanent Critique of Race in Science

As we can observe, not contextualization per se but the choice of context leads to reductionism. One of the shared features of Blumenbach's historicizations, inevitably elided into a simplified vision, is what White has defined as a "cold" chronology, with fewer dates and events than a "hot" one (White ibid 24). Minimalist and short-term retrospectives of Blumenbach's legacy have reverberated with no less problematic traditions to emancipate the contemporary fight against scientific racism from its "dark" past.

We share a critical view on such transcendent critique as obsolete (Feder, 2016), which treats the issue of science in one-size-fits-all manner (La Capra ibid). A more inherently dialectical, immanent critique that avoids speaking the language of false escape (Feder ibid) invests in what Dominik LaCapra defines as "self-critical historiography," aimed at reflecting upon our own interrelation with issues such as race in science and epistemic injustice. But is a long-term or "hot" chronology an exhasuting precondition for overcoming the weaknesses of reductionist contextualism?

For instance, the historicization of embryology, intertwined with systemic racism and systemic sexism, follows the ongoing reproduction of the great chain of being as a cross-temporal complex metaphor through an impressively long timeline (Gilbert, 2021). The generations of analogies aimed at "proving" a hierarchical order of social life with women and non-whites occupying lower positions because they are considered not fully developed embryos of white men. While this approach provides a nuanced understanding of the intersectionality between gender and race, the historical predispositions for emancipating knowledge in embryology from systematic injustice are missed. Instead, Gilbert focuses on the driving forces behind the ongoing reproduction of the great chain of being as a "Zombie idea," which prevents the rejection of knowledge based upon superstition, even in the face of empirical evidence. This idea builds Gilbert's historicization in the tradition of transcendent critique that does not furnish the methods for reflecting upon the complexity of the relationship between pre-modern and modern epistemologies. In short, it offers a stringent picture of how science should *not* be done but lacks specification for identifying desirable directions in knowledge production. It further questions the call for revising the agenda of embryology articulated by Gilbert himself, who reconsidered the options for and limits of reintroducing the agenda of developmental organicism in the face of extensive changes in genetics, ecology and modeling in developmental biology (Gilbert & Sarkar, 2000).

Operating as a commitment to a social, rather than to a moral or a dogmatic, critique, immanent critique is embedded in particular communities, including camps of scientists whose members reflect their standards, often leading to demands for changes in these standards (Stahl, 2013). As a promising strategy for transforming normative standards, including knowledge production and its implications, immanent critique is driven by the contradictions between ideology and reality (Antonio, 1981 338). The power and authority of such critique, rooted in collective reflections, reveals the continuities and changes in normative standards, as well as explanations for these processes—in short, its historicization.

Hans Blumenberg insists that investigations into the signified that frame the longue durée of epistemologies, like truth metaphors, can be proceeded by placing *longitudinal sections,* which emphasize the deficiency of historical material as a challenge to the narrator. This initial self-critique operates in favor of equipping the narrator with a series of points (unknown, not explored, or unfamiliar events) through which a curve of critical historical narrative may be drawn (Blumenberg, 2010, 31). In this turn, Gilbert's narrative provides a thoughtful problematization of the vicissitudes of embryology through premodern and modern eras—including the puzzling question of why influential embryologists like Johann Ferdinand Autenrieth,[2] who followed Blumenbach's pathway and who were abolitionists, are absent in such narration.

Accepting the challenge of missing events and historical facts calls for reorganizing approaches to timelines of narration. Sylvia Wynter's differentiation between genealogy (one counter exertion as the historical-structural route to the other) and parallelism (one counter exertion mirroring another) (Garba Sorentino, 2022, 8) as two interrelated practices of critical historicization can be seen as a promising direction for developing an immanent critical approach. Understanding Blumenbach's affinity with previous cohorts of scholars and those who have followed in terms of genealogy and parallelism becomes even more demandable if we add (non)whiteness to the well-established discussion regarding race, species, and variation (Hudson, 1996).

The ambitions of Blumenbach indeed aimed at exploring larger populations by "subject[ing] the human species to the same kind of biological analysis and classification previously used only for plants and animal" (ibid., 252). With this exceptional contextualization, the Blumenbachian idea of human variation, a/k/a the reflection of plasticity, sharpens his agreement with Buffon and the opposition to Linné as Hudson offers. But claiming blurred boundaries between the categories of race and species seems to be an insufficient explanation for exploring the trend to indicate that exceptional human groups belonging to humanity (in terms of the general direction

[2] Johann Ferdinand Autenrieth (1772–1835) one of the pioneers of forensic medicine, the author of *Supplementa ad Historiam Embryonis Humani* (1797) written in cooperation with Blumenbach. During his one year stay in the United States Autenrieth supplied Blumenbach with a solid portion of knowledge about the northern Indians and samples of their skulls.

of human progress) is seen as questionable because of their degradation, as among Lapps and other Arctic peoples, in Buffon's view.[3]

The motif of moving degradation to total extinction is one of the most echoed continuities in race-informed constructions of human groups from pre-modernity through modernity, and until nowadays. Darwin, who accepted Blumenbach's idea of transitoriness of human "races," articulated the multifold threat of extinction which "follows chiefly from the competition of tribe with tribe, and race with race" and recognized "the grade of their civilisation [as] to be a most important element in the success of competing nations" (Darwin, 1871). In the middle of the twentieth century, Ruth Benedict had stressed the tragic destiny of the "pygmy" and "semi-pygmy" peoples due to their smaller brain size and the questionable ability among these groups to function in modern civilization. Remarkably, in discussing this "purely academic question," Ruth Benedict contrasts these groups to the already disappeared Bushmen who "might not have been able to adjust themselves to machine civilization, but in spite of their cranial capacity they were not a reproach to the human race." (Benedict & Mead, 2019, 50 51) One decade later, Joseph Sydney Weiner, one of the ideological inspirers of the International Biological Program, which connected racially minded experts across the globe during the 1960s and 1970s, brought forward the mission of exploring genetically important "socially" or linguistically isolated groups in industrialized communities until they disappeared (Weiner, 1964, 191 193).

Following the construction of blackness through the "symbolic death role" of Black people (Garba & Sorentino, 2022) and non-whiteness as a dead-end option introduces new connections between pre-modernity, with its primary focus on salvation, and the modern obsession with progress and development. Blackness as "one of material-conceptual indeterminacy mediating the discontinuity of epochs" (ibid) clearly determines the Blumenbachian vision on non-white humans and their destiny.

Theorizing the difference between whites and people of color in terms of two extremes regarding human maturity had inclined Blumenbach to situate other races or varieties as transitioning between Caucasian and Africans. This construction unites pre-modern focus on salvation with modern idea of progress as "something that results from an immanent process of development." (Turner, 2020, 135) In this sense, the abolitionist sentiment shared by Blumenbach stemmed from considering slavery as a "deadening" practice that prevented the desirable scenario of redemption for non-whites. The fact that Blumenbach's whiteness was first biological and then cultural, and it was further associated not with Europeanness but with the cradle of human civilization, only intensified this oscillation between pre-modern metaphor and modern rationalization.

This pre-modern/modern duplicity of the Blumenbachian approach to human variation reverberates with Blumenberg's understanding of the meaning of historicization as a transfer from anxiety in front of unknown and not yet examined to a

[3] The partial explanation for Buffon's attention to those designated as "endangered species" can be found in Schmitt, Stéphane. "Buffon's theories of generation and the changing dialectics of molds and molecules." In *Philosophy of Biology before Biology*, 27–47. Routledge, 2019. Schmitt stresses the increased role of the historical approach in the transformation of Buffon's epistemology.

more bearable and manageable fear (ibid 132). Through this implicative connection, the relationship of metaphorology, which emblematizes pre-modernity, to the history of concepts (in the narrower, terminological or even modern sense) is defined as "an ancillary one: metaphorology seeks to burrow down to the substructure of thought, the underground, the nutrient solution of systematic crystallizations; but it also aims to show with what 'courage' the mind preempts itself in its images, and how its history is projected in the courage of its conjectures." (Blumenberg, 2010, 5).

Blumenbach's courage in telling the history of humanity as a transition from non-whiteness to whiteness is a part of story about transformation as a fundamental alteration of state (Few et al., 2017). In Blumenberg's understanding, it should dispel the fear among Blumenbach's contemporaries in the face of unknown humans and their presence in a human past and present. In Blumenbach's narrative, the shift from the category of evolutionary outcomes such as species or race to processes such as adaptation, transformation, or development is stabilized by the omnipresent whiteness as the only desirable pattern of such transformation. Humans who were, to Blumenbach, champions in transformability, deserve this option for salvation. Further, in Blumenbach's methods for amassing knowledge, it is easy to recognize the "modern" mode of acquiring such knowledge, which Blumenberg examined through the metaphor of "an act of violence" (Blumenberg ibid 7) that distinguished modern epistemologies.

The transformative drive introduced by Blumenbach as a universal feature of all humans can be easily interpreted as bridging the myth of human salvation and a purely biological approach to explaining human variation. The mythological component of the transformative drive helps Blumenbach to place the environment at the margins of his narration and to focus on the temporal measurement of human variation. This turn had shaped the abiding agenda of examining the risks of human extinction in terms of race through Darwinian modeling of the evolutionary process (Beer, 2009) to later conceptualizations of adaptability and more recent interdisciplinary research on the role of human activity (Baumsteiger & Moyle, 2017).

The intensive use of visualization in Blumenbach's texts aimed at discussing human variation can be interpreted as a reflection of his involvement in the "modern project of building a neutral and geometric space of meaning, free from all difference." (Petteni, 2020) Moreover, the richness of Daniel Chodowiecki's illustrations for Blumenbach's *Beyträge zur Naturgeschichte* (1790) and the choice of the illustration for *Abbildungen Naturhistorischer Gegenständen* (1796) help to form a consistent application of the multiple metaphors of (im)maturity for the races.

Wynter, who applied Blumenberg's approach to connecting pre-modernity and modernity for constructing the idea of race as preoccupation, recognizes two modes of race, premodern, race-as-substance, and modern, race-as-political-function (Wynter, 2003). The Blumenbachian transitory vision of human variation is aligned with the pre-modern idea of race as substance, as well as Darwin's interpretation of human variety and its origins. With his abolitionist claims, Blumenbach had shifted to operating with race as a political function. Being aligned with the Enlightenment view on natural law as having a dual character, namely, a physical and a moral character (i.e., related to human actions), Blumenbach deepens this stance through claiming historical interrelation among human variations (De Angelis, 2024). With this shift, Blumenbach engages the game of legitimacy and the appropriation of control over human lives, two emblematic features of modernity. In this context, the contemporary historicization of Blumenbach's legacy reverberates with the challenges Blumenbach himself faced.

1.3 Blumenbach's Legacy in the Lens of Adaptation

Despite long-term reductionist contextualism, and partially because of it, Blumenbach has attained a unique recognizability among further cohorts of anthropologists. The contributors of this volume unpack this "generally circulated cultural memory" (Ellis, 1982) through investigating different "adaptations" of Blumenbach's texts. There are two main reasons beyond our choice for exploring Blumenbach's legacy through the lens of adaptation. Firstly, according to Linda Hutcheon, exploring adaptation brings forward "the urge to adapt and therefore perhaps the very process of adaptation." (Hutcheon, 2006, 107). Understanding why one or another text continues to be adapted implements tracing "the registering of human consciousness in the text itself," (ibid) and our contributors solve this task regarding the adaptation of Blumenbach in different contexts.

This historical variety explains the other reason for choosing adaptation as a method for examining Blumenbach's legacy, namely, sharpening "the possible conflict between chronologically and syntactically structured series [for us, cases of adaptation]." (White ibid., 25) Bringing together various adaptations is decisive, not only for questioning already established historical imperatives but also for practicing historicization through Wynter's concepts of genealogy and parallelism. Further, Blumenbach's texts (in the wider meaning meant to include his way of visualizing races) have become canonical to such a degree that many readers of adaptations have no direct experience of Blumenbach's texts. This prompts us to trace Blumenbach's adaptations, especially those that turned the sacred elements of his approach to human variation on their head.

At the center of the chapter by Simone De Angelis, we can locate the adaptation implemented by Blumenbach himself, who sought to tell the story of humanity as part of a general natural history, but with an emphasis on the changing role of "man" in the vicissitudes of humanity. ***"In The Course of Time": The Formative Drive and "Dégénération" in Johann Friedrich Blumenbach's Varieties of Humankind*** revises the role of the relationship between Blumenbach and Kant, moving toward distinguishing the uniqueness and commonness of Blumenbach's vision. It navigates a detailed analysis of intertwining geology and biology for temporalizing humanity. De Angelis efficiently shows how instructive theological argument (and metaphor, including the "great chain of being") has been for developing modern methodology of historicizing.

Victoria Shmidt situates adaptations of Blumenbach by following generations of thinkers in the context of colonial power and its tectonic transformation determined by delegitimizing slavery. In her chapter, ***Blumenbach's Conceptualization of Human Perfection and its Evolution through the History of Colonial Power***, Blumenbach's texts are explored as one of the starting points for long-term patterns that appropriate initially anti-racist arguments by racially minded scholars. To Shmidt, this upwardly mobile cultural accreditation crucial for race in science has helped "to gain respectability or increase cultural capital." (Hutcheon ibid 91).

Exploiting the best options for historical comparison, Malin Sonja Wilckens demonstrates how another instrument of adaptation, namely, "repetition with variation" assisted U.S. and French anthropologists, adherents of polygenism, helped to transform Blumenbach's approach to human variation. ***The Dissemination and Alteration of Blumenbach's Racial Classification in Nineteenth-Century Polygenetic US-Ethnology and French Anthropology*** sheds light on the features of Blumenbach's texts which have made their further racialization possible. Wilckens problematizes the recent dogmatic critique of Blumenbach as one of those who pioneered the long line of skull collection that led to the "epidemic" of racially based comparisons. In her interpretation, a transnational community of overtly racist anthropologists (who followed Blumenbach in his obsession with collecting human skulls for proving the polygenetic, rather than the monogenetic, origins of human diversity) achieved long-term recognition by recording purely applied research, alienated from much of the epistemic nuances of variety and race.

Ivana Prazic, Khairunnisa A Shukor, Sara Shakilla Mohd Salim, and Shakila Che Dahalan interrogate the invention of the Malay variety by Blumenbach as an exemplary case of interdisciplinary racialization, in response to the call for such research introduced in the chapter by De Angelis. ***Blumenbach's Malay Variety: Language and Emotionality in a Transitory Race*** not only deepens the newly established trend of exploring the Blumenbachian approach from the periphery (Kühnast, 2020, 56) but also brings in a feminist optic for emancipating the idea of a united national group from race-informed argument. Prazic and her team move far beyond a well-established historicization of Blumenbach's approach in non-European contexts as dissemination (Demel, 2015; Jiang, 2012), deconstructing Blumenbach's concept of the "Malay variety" as transgenerational phenomena reflected in the educational curriculum.

Cosmin Koszor-Codrea turns the reader's attention to the acceptance of Blumenbach in Romania and introduces the complex political and social contexts for clarifying how multidimensional the influence of Blumenbach's approach to human variation was in shaping racial hierarchies. Koszor-Codrea links the racialization of Roma and people of African descent through bringing into focus the case of Romanian slavery. The chapter resonates closely with debates about the compensation of slaveowners in the British Caribbean after abolition in 1833 (Hall, 2020). Like the authors of the chapter about the Malay variety, ***Pedagogies of White Domination: Blumenbach's Legacy and Racial Classifications in Romanian Science and Secondary Schools, 1870–1914*** connects the reproduction of Blumenbach's approach and education as a main trajectory for fixing race-informed visions on non-whites.

Kseniya Brailovskaya moves from the implicit presence of Blumenbach's concept of the "driving force" in the history of scientizing congenital malformation, typical of recent research (Rich, 2023), to follow a double process of "interpreting and then creating something new." (Hutcheon ibid 20) She does so through the critical reconstruction of Russian, and then Soviet, conceptualizations of oligophrenia—an umbrella concept applied for exploring mental "deficiency." ***Imperfect Evolution: A Formative Force in Soviet Studies of Oligophrenia*** contrasts the straight interpretation of intersectionality among gender, disability, and race as producing multiple analogies regarding inferiority, in favor of a more nuanced historicization of Russian special education. Brailovskaya excavates historical facts regarding the adaptation of Blumenbach by scholars of German origin who practically built the basis of Russian special education, along with a wide range of metaphors introduced by this cohort into the longue durée of connecting human perfection with harmony and its changing signifiers.

The adaptation of Blumenbach's legacy has taken place in contexts, but alongside the many instances of Blumenbach's adaptation, there has been (and continues to be) an exchange or circulation of approaches. The entangled history of their interrelationships remain an unanswered challenge to our understanding of how racial science is not only reproduced but also legitimized. Since Blumenbach invented his five types of human varieties, scientists around the globe have introduced about 2,000 classifications based on race-informed categories into the field of anthropology. These classifications have directly influenced population politics, public health, and social protection. Moreover, racial classifications have shaped arguments justifying the surveillance of population groups, not uncommon in extermination policies. At the same time, racial classifications, to a greater or lesser extent, have reflected the formation of knowledge about the differences among people and human sameness. Criticism of the way racial classifications have been applied remains detached from recognizing progress in explaining human differences and similarities.

Blumenbach, whose scientific contribution consists of synthesizing and refining already invented ideas and principles (which excludes him from historical narratives about innovators like Buffon or Kant), has emerged as one of the pioneers of a modern, politicized science. This is disconcerting and often leads to the excessive pathos in historicizing Blumenbach's legacy that overprotects against the need to link the critical narrative with recognition. The contributors of this volume show how fruitful the work on such challenges can be.

References

Ankersmit, F. R. (1989). Historiography and postmodernism. *History and theory*, *28*(2), 137–153.

Antonio, R. J. (1981). Immanent critique as the core of critical theory: Its origins and developments in Hegel, Marx and contemporary thought. *British Journal of Sociology*, 330–345.

Baumsteiger, J., & Moyle, P. B. (2017). Assessing extinction. *BioScience, 67*(4), 357–366.
Beer, G. (2009). Darwin and the uses of extinction. *Victorian Studies*, 321–331.
Benedict, R., & Mead, M. (2019). *Race: Science and politics*. University of Georgia Press.
Blumenbach Johann Friedrich to Johann Heinrich Blumenbach in Gotha Göttingen, ca. 30 October 1777, Letter 70. In F. W. P. Dougherty (Ed.), *The correspondence of Johann Friedrich Blumenbach 1773–1782*, (vol. 1, pp. 1–230), Göttingen: 2006
Blumenberg, H. (2010). *Paradigms for a metaphorology* (p. 31). Cornell University Press.
Darwin, C. (1871). *The descent of man, and selection in relation to sex*. Printed by William Clowes and Sons.
De Angelis, S. (2024). Das Menschenbild der, Wissenschaft vom Menschen' (1660–1800). In M. Zichy (Ed.), *Handbuch Menschenbilder* (pp. 33–49). Springer Fachmedien Wiesbaden.
Demel, W. (2015). Constructing racial theories on East Asians as a transnational "Western" enterprise, 1750–1850. In *Race and racism in modern East Asia* (pp. 55–76). Brill.
Depew, D. J. (2017). Darwinism in the twentieth century: Productive encounters with saltation, acquired characteristics, and development. In R. G. Delisle (Ed.), *The Darwinian tradition in context: Research programs in evolutionary biology* (pp. 61–88). Springer International Publishing.
Depew, D. J., & Weber, B. H. (2011). The fate of darwinism: Evolution after the modern synthesis. *Biological Theory, 6*, 89–102.
Eberl, O. (2019). Kant on race and barbarism: Towards a more complex view on racism and anti-colonialism in Kant. *Kantian Review, 24*(3), 385–413.
Ellis, J. (1982). The literary adaptation. *Screen 23*(1), 3–5. 3.
Fabian, J. (2014). *Time and the other: How anthropology makes its object*. Columbia University Press.
Feder, H. (2016). *Ecocriticism and the idea of culture: Biology and the bildungsroman*. Routledge.
Few, R., Morchain, D., Spear, D., Mensah, A., & Bendapudi, R. (2017). Transformation, adaptation and development: Relating concepts to practice. *Palgrave Communications, 3*(1), 1–9.
Flaherty, G. (1989). Goethe and shamanism. *MLN, 104*(3), 580–596.
Frye, N. (1961). Myth, fiction, and displacement. *Daedalus, 90*(3), 587–605.
Garba, T. P., & Sorentino, S. M. (2025). Blackness before race and race as reoccupation: Reading Sylvia Wynter with Hans Blumenberg. *Political Theology*. 1–22. 8
Gilbert, S. F. (2021). Systemic racism, systemic sexism, and the embryological enterprise. *Developmental Biology, 473*, 97–104.
Gilbert, S. F., & Sarkar, S. (2000). Embracing complexity: Organicism for the 21st century. *Developmental Dynamics: An Official Publication of the American Association of Anatomists, 219*(1), 1–9.
Gould, S. J. (1977). *Ontogeny and phylogeny* (Vol. 9). Belknap Press of Harvard University Press.
Gould, S. J. (1997). *Darwinian fundamentalism*. New York Review of Books.
Gould, S. J. (2014). The geometer of race. In *The concept of race in natural and social science* (pp. 1–5). Routledge.
Goy, I. (2017). *Kants Theorie der Biologie: Ein Kommentar. Eine Lesart. Eine historische Einordnung*, De Gruyter.
Hall, C. (2020). The slavery business and the making of "race" in Britain and the Caribbean. *Current Anthropology, 61*(22), 172–182.
Herbert, S. (2005). The darwinian revolution revisited. *Journal of the History of Biology, 38*, 51–66.
Hill, J. D. (2017). Is it moral to hold a racial identity? A cosmopolitan response. In N. Zack (Ed.),*Oxford Handbook of Philosophy and Race* (pp. 402–410).
Hill, J. D. (2021). *What do white Americans owe black people?: Racial justice in the age of post-oppression*. Emancipation Books.
Hudson, N. (1996). From "nation" to" race": The origin of racial classification in eighteenth-century thought. *Eighteenth-Century Studies, 29*(3), 247–264.
Huneman, P. (Ed.). (2007). *Understanding purpose: Kant and the philosophy of biology* (Vol. 8). University Rochester Press.

Hutcheon, L. (2006). *A theory of adaptation*. Routledge.
Ingram, D. (1990). Blumenberg and the philosophical grounds of historiography. *History and Theory, 29*(1), 1–15.
Jiang, S. (2012). Blumenbach in East Asia: The dissemination of the "five-race theory" in East Asia and a textual comparison. *Oriens Extremus, 51*, 107–153.
Junker, T. (1998). Blumenbach's racial geometry. *Isis, 89*(3), 498–501.
Junker, T. (2018). Blumenbach's theory of human races and the natural unity of humankind. In *Johann Friedrich Blumenbach* (pp. 96–112). Routledge.
Kansteiner, W. (1993). Hayden White's critique of the writing of History. *History and Theory, 32*(3), 273–295. https://doi.org/10.2307/2505526
Keel, T. (2018). Impure thoughts: Johann Blumenbach and the birth of racial science, In *Divine variations: How christian thought became racial Science*. Redwood City.
Kühnast, A. (2020). Johann Friedrich Blumenbach's 'Neuholländer'. *Australian Studies Journal Zeitschrift für Australienstudien 33/34*, 31–56.
LaCapra, D. (2013). *History, literature, critical theory*. Cornell University Press.
Lambier, J. D. (2016). Romantic evolutions: Introduction. *Literature Compass, 13*(10), 587–596.
Lenoir, T. (1989). *The strategy of life: Teleology and mechanics in nineteenth-century German biology* (Vol. 13, p. 9). University of Chicago Press.
Mareike Knoke Sensible Provenienzen—Die Spur der Gebeine (2023). https://www.volkswagenstiftung.de/de/news/story/sensible-provenienzen-die-spur-der-gebeine
Marino, M. (2022). At the roots of racial classification: Theory and iconography in the work and legacy of Johann Friedrich Blumenbach. In *Encyclopédie internationale des histoires de l'anthropologie* (Available online https://www.berose.fr/article2546.html?lang=fr)
Mazzolini, R. G. (2018). A defense of human rights: Blumenbach on albinism. In N. Rupke, & G. Lauer (Eds.), *Johann Friedrich Blumenbach* (pp. 113–120). Routledge.
Mills, C. W. (1997). *The racial contract*. Cornell University Press.
Niekerk, C. (2014, November). Buffon, Blumenbach, Lichtenberg, and the origins of modern anthropology Talk—Workshop. In The nature of man: Norms and deviations in the 18th century (pp. 12–14). Lichtenberg-Kolleg, Georg-August-Universität Göttingen.
Outram, D. (1986). Uncertain legislator: Georges Cuvier's laws of nature in their intellectual context. *Journal of the History of Biology, 19*, 323–368.
Petteni, O. (2020). The ideal of optics and the opacity of life: Blumenberg on modernity and myth. In *Interrogating modernity: Debates with Hans Blumenberg* (pp. 193–213).
Raj, B. (2007, December 22). The beautiful skull and Blumenbach's errors: The birth of the scientific concept of race. *BMJ. 335*(7633), 1308–9. https://doi.org/10.1136/bmj.39413.463958.80. PMID: 18156242; PMCID: PMC2151154.
Rambachan, A. (2018). Overcoming the racial hierarchy: The history and medical consequences of "caucasian." *Journal of Racial and Ethnic Health Disparities, 5*(5), 907–912.
Rich, M. (2023). Monstrosity in medical Science: Race-making and teratology in the nineteenth-century United States. *Isis, 114*(3), 513–536.
Richard, M. (2004). Contextualism and relativism. *Philosophical Studies: An International Journal for Philosophy in the Analytic Tradition 119*(1/2), 215–242.
Rupke, N. (2018). The origins of scientific racism and Huxley's rule. In *Johann Friedrich Blumenbach* (pp. 233–247). Routledge.
Rüsen, J. (1982). Die vier Typen des historischen Erzählens, In Reinhart Koselleck/Heinrich Lutz/ Jörn Rüsen (Hg.), *Formen der Geschichtsschreibung*, Münchenf. S. 514–605
Rüsen, J. (2005). *History: Narration, interpretation, orientation*. Berghahn Books.
Schmidt, V. (2014). *Global modernity: A conceptual sketch*. Springer.
Sloan, P. (2019). Metaphysics and "vital" materialism: Émilie du Châtelet and the origins of French vitalism. In *Philosophy of biology before biology* (pp. 48–65). Routledge.
Stahl, T. (2013). *What is Immanent critique?* (SSRN Working Papers). http://ssrn.com/abstract=2357957
Táíwò, O. (2022). *Against decolonisation: Taking African agency seriously*. Hurst & Co

Turner, Z. (2020). Talay. Political legitimacy and founding Myths. In *Interrogating modernity: Debates with Hans Blumenberg* (pp. 129–147).
Watson, M. C. (2011). Cosmopolitics and the subaltern: Problematizing Latour's idea of the commons. *Theory, Culture & Society, 28*(3), 55–79.
Weiner, J. S. (1964). The biology of man in the international biological programme. The human adaptability project current Anthropology, *5*(3), 191–193.
White, H. (1998). The historical text as literary artifact. In B. Fay, P. Pomper, & R. T. Vann (Eds.), *History and theory contemporary readings* (vol. 22). Blackwell Publishers.
Wolfe, C. T. (2014). Teleomechanism redux? Functional physiology and hybrid models of life in early modern natural philosophy. *Gesnerus, 71*(2), 290–307.
Woodard, B. (2020). Outgrown purpose, outlived use: On parasitic teleology. In M. Rosen (Ed.), *Diseases of the head,* (pp. 43–70). Punctum Books.
Wynter, S. (2003). Unsettling the coloniality of being/power/truth/freedom: Towards the human, after man, its overrepresentation—An argument. *CR: The new centennial review* 3(3), 257–337.
Zammito, J. H. (2018). *The gestation of German biology: Philosophy and physiology from Stahl to Schelling*. University of Chicago Press.

Victoria Shmidt is senior postdoc researcher at the Centre for History of Science at the University of Graz, Austria and principle investigator of the Project A Critical History of Genetics and its Implication in Central Eastern Europe supported by FWF. Her recent publications focus on race in science and its operation in Central Europe: "Historicizing Roma in Central Europe Between Critical Whiteness and Epistemic Injustice" (Routledge, 2021); "A Critical History of Health Films in Central and Eastern Europe and Beyond" (Routledge, 2023); special issue of Romani Studies "Epistemologies in Romani studies: Moving beyond othering otherness" (2024).

Simone De Angelis is Professor of History of Science and Director of the Centre for History of Science at the University of Graz, Austria. His main research areas are Renaissance science and early modern history and philosophy of science and medicine, Enlightenment studies, history of life sciences, and anthropology. He has been Fellow of the Swiss National Science Foundation, the Herzog August Bibliothek in Wolfenbüttel, Visiting Researcher at the Wellcome Trust Centre for the History of Medicine at University College London and the Università degli studi di Padova. He is a Visiting Professor at École des Hautes Études en Sciences Sociales (EHESS) and the Centre Alexandre Koyré (CAK) in Paris. His books include Von Newton zu Haller (Niemeyer 2003), Scholarly Knowledge. Textbooks in early modern Europe (Droz 2008), Anthropologien (De Gruyter 2019, 2nd edition).

Chapter 2
"In the Course of Time". The Formative Drive and "Dégénération" in Johann Friedrich Blumenbach's Varieties of Humankind

Simone De Angelis

Abstract This chapter examines Blumenbach's writing on the varieties of the human race in the context of the Blumenbach-Kant debate around 1790 concerning the idea of the formative drive. This examination primarily involves demonstrating how the concept of "dégénération" is linked to a new understanding of natural history in terms of the historicisation of nature. The historicisation of nature is linked to a late eighteenth century subversion of the classical idea of the "chain of beings," which dissolves, as it were, within the framework of a temporal axis and thus becomes temporalised. The natural history of the Enlightenment is discussed with regard to Blumenbach as follows: (1) *dégénération* as a theory of migration and progress; (2) the naturalisation of "man" and the notion of the "ignoble savage"; (3) the notion of the formative drive and a new archaeology of the globe; (4) the formative drive and *dégénération*; (5) geographical history and the comparison of human faces and skulls; and (6) *Varietas Americana*. The central goal is the inclusion of Blumenbach's natural history into the natural law tradition of the Enlightenment.

Keywords Formative drive · *Dégénération* · Natural history · Temporalisation · Naturalisation of history · Archaeology of the globe · Geographical history · Varieties of humankind · Natural law

2.1 Introduction

For an adequate understanding of Blumenbach's concept of *dégénération* in *De generis humani varietate nativa* (1795, 3rd edition), it is crucial to consider the shifts in natural history since 1750. Kant's idea of natural history as an "archaeology of nature" and Blumenbach's model of Earth's revolutions (upon which Kant draws)

S. De Angelis (✉)
Centre for the History of Science, University of Graz, Graz, Austria
e-mail: simone.de-angelis@uni-graz.at

V. Shmidt and S. De Angelis (eds.), *Blumenbach's Racial Classification*, History, Philosophy and Theory of the Life Sciences 39,
https://doi.org/10.1007/978-3-032-15950-2_2

are centrally relevant in this context. Blumenbach's focus on fossils, and his comparative method, which relies on precise observation, inherently also rejects Buffon's theory of types, which the Göttingen naturalist must have deemed too speculative. In light of a new conception of natural history around 1790 and against the backdrop of his concept of the formative drive (*Bildungstrieb*), Blumenbach did not consider *dégénération* as a deviation from an "original primitive type."[1] For Blumenbach *dégénération* instead signifies a transformation of, or deviation from, the formative drive which, in the context of Earth's historical changes, coincides with the revolutions in nature and subsequently yields the varieties of humankind ("*degeneratione in varietate*") (Blumenbach, 1795, § 33, 88). Blumenbach's concept of *dégénération* is thus an approach for addressing the issues of temporality and transformations in nature. That is to say that *dégénération* is a positive concept, a means of understanding the specifics of nature's transformations that goes hand in hand with the idea of temporalisation, which also implies that the emergence and classification of human varieties or races (*Racen*) can primarily be understood as a function of time. Moreover, in the third edition of *De generis humani varietate nativa*, Blumenbach draws on Kant's integration of the two explanatory models for the formation of organic bodies (the physical–mechanical and the teleological) (Blumenbach, 1795 § 33, 82–83). It is therefore important to understand and further contextualise Blumenbach's treatise on the varieties of humankind from the perspective of his work on the formative drive. Central aspects of Blumenbach's *Über den Bildungstrieb* (1789, 2nd edition) and the reception of the formative drive concept in Kant's *Critique of the Power of Judgment* (1790) have already been examined in detail elsewhere (De Angelis forthcoming). This chapter aims to more thoroughly historicise Blumenbach's *De generis humani varietate nativa* (1795, 3rd edition) and contextualise it within the Enlightenment debate on natural history. This primarily involves demonstrating how the concept of *dégénération* is linked to a new understanding of natural history in terms of the historicisation of nature. The historicisation of nature is linked to a late eighteenth-century subversion of the classical idea of the "chain of beings," which dissolves, as it were, within the framework of a temporal axis and thus becomes temporalised. After all, the eighteenth century is neither aware of evolution sensu Darwin, nor is the ca. 1800 idea of the transformation of nature "pre-Darwinian". This transformatory notion rather constitutes a case of historicising the natural process and changes the concept of natural history in the works of Blumenbach and Kant at the end of the eighteenth century. To historicise Blumenbach's text, moreover, implies drawing on the original Latin text as well as reconstructing and thoroughly examining key terms throughout the various sources and based on the references in the footnotes.[2] The natural history of the Enlightenment ultimately encompasses several interconnected elements and dimensions, which can be examined in relation to Blumenbach

[1] Doron (2016, 383), for example, claims that, according to Blumenbach, "the races" are merely "degenerations of an original common type" ["*les dégénérations d'un type originel commun*"].

[2] From a philological perspective, it is essential to consult the original Latin text of the third edition of Blumenbach's *On the Varieties of Humankind* (1795). Thomas Bendyshe's 1865 English translation reflects the polygenist and racist convictions of both the translator and the Anthropological Society of London, of which he was a member. See Blumenbach 2014 [1790], editor's introduction, XI–XIII.

as follows: (1) *dégénération* as a theory of migration and progress; (2) the naturalisation of "man" and the notion of the "ignoble savage"; (3) the notion of the formative drive and a new archaeology of the globe; (4) the formative drive and *dégénération*; (5) geographical history and the comparison of human faces and skulls; (5) *Varietas Americana.* To conclude, I briefly contextualise Blumenbach's natural history within the Enlightenment tradition of natural law, which ultimately forms the bedrock not only for his commitment to the unity and equality of humankind but also notably for his strong critique of the slave trade at the end of the eighteenth century.

2.2 *Dégénération* as a Theory of Migration and Progress

Like many eighteenth-century natural scientists and authors, Blumenbach was confronted with questions regarding the age of the Earth and the chronology of human history. In *Beyträge zur Naturgeschichte* (1790), Blumenbach located the revolutions of the Earth within a pre-Adamite world that far exceeded the biblical chronology since creation (5,000–6,000 years ago). He also spoke of an initial "total revolution" of the Earth, during which the formative drive changed course, so that newly emerged living beings differed from those of the first creation and continued to evolve through subsequent revolutions. Thus, Blumenbach essentially adhered to a biblical notion, even though he clearly broke with its timeframe, extending it with regard to the pre-Adamite world. The eighteenth century, in fact, already acknowledged this temporal extension: long before the rise of geology and prehistory in the nineteenth century, French consul Benoît de Maillet (1656–1738), for example, proposed in his clandestinely circulated text *Telliamed* (1755) a hypothesis of the sea's gradual retreat and the origin of life in water, estimating that humans had inhabited the Earth for over 500,000 years (de Maillet, 2023, 231–232). This hypothesis highlights a chronology-related problem within "pre-archaeological scholarship." As intellectual historian John G. A. Pocock puts it, we should be cautious about viewing the Book of Genesis, in line with nineteenth-century intellectual history, as an obstacle to the development of a "scientific anthropology," not least because the eighteenth century was "an age before archaeology, when neither the science of geology, the origin of species, nor the discovery of fossil man had yet raised problems insoluble through reliance on the written record of early civilisation" (Pocock, 2009, 44). Antoine-Yves Goguet (1716–1758), a member of the Paris Parliament, faced the same problematic. In his work *De l'Origine des Loix, des Arts et des Sciences* (1758), he outlined a history of material culture and intellectual development that remained within the biblical framework. However, this did not prevent him from developing "Enlightened schemes of human progress" (Pocock, 2009, 44). Early on, in the preface, Goguet states: "The history of laws, techniques, and sciences is, strictly speaking, the history of the human mind" (Goguet, 1759, Preface, V). Goguet's sources for his theory of progress included the Book of Genesis as well as ancient texts by Homer, Hesiod, Herodotus, Diodorus, Vitruvius, Strabo, Seneca, Pliny, and Plutarch, all of whom he regarded as witnesses in terms of the ancient and late medieval doctrine of authority and testimony (Danneberg, 2009; De Angelis, 2013). Against this backdrop, for Goguet, the history of humankind commences approximately 150 years after Noah and the flood, i.e., after Noah's descendants settled on the Plain of Shinar, amply procreated, and started to spread across and repopulate different parts of the world (Goguet, 1759, Introduction, 1–2).[3] This scenario represents the onset of the development of *dégénération* into a theory of migration; the families dispersed from the Plain of Shinar to all parts of the world, languages diversified, and humanity—following the unsuccessful attempt to maintain unity and a common language in the Tower of Babel episode—became scattered (Goguet, 1759, Introduction, 3–4). Furthermore, Goguet links *dégénération* to a theory of knowledge, suggesting that this dispersion of humankind ("*transmigrations*") obscured humanity's primordial knowledge ("*des connoissances primitives*") (Goguet, 1759, Introduction, 6): the memory of pre-catastrophic knowledge, if not entirely lost, became profoundly defaced and obscured ("*altérée & obscurcie*") (Goguet, 1759, Preface, XX). This concept of "primordial knowledge," here adapted to reflect humanity's history after the Fall, structurally parallels the philosophical–theological concept of natural knowledge (*notitiae naturales*), implanted in the human mind through creation but clouded by the Fall (De Angelis, 2004). This perspective finally forms the basis for *dégénération* as a theory of civilisation and "savagery" or "barbarism": The diversity of languages and the isolation of families led to deep ignorance ("*ignorance*"), plunging the world into extreme "barbarism" ("*barbarie extreme*"): "One could see men wandering in the woods and fields, without laws, without order, without leaders. Their ferocity became so great, that some started to devour each other"; they even forgot the most basic skills such as the use of fire, so that the first humans were barely distinguishable from animals in their way of life (Goguet, 1759, Introduction, 6–7). The underlying argumentative structure is the key element here: It is based on the comparison between the "first humans" ("*les premiers hommes*") and the so-called "savage peoples" ("*les Sauvages*"), primarily the Americans ("*Americains*") as described in contemporary travel literature ("*relations modernes*") (Goguet, 1759, Introduction, 8). This method of comparison had been introduced by Jesuit missionary Joseph-François Lafitau (1681–1746) in his work *Mœurs des Sauvages Amériquains, Comparées aux Mœurs des Premiers Temps* (1724). Goguet focusses on the correlation between mobility and "savagery"—he notes that "savages", Tartars, and Arabs, still travel long distances with their families—and on the association of "first humans" with "savages." This nomadism is what led to the dehumanisation of South and Central American peoples, described as "void of all principles of humanity" ("*dénués de tous les principes de l'humanité*") (Goguet, 1759, Introduction, 5, 8–9). This attitude found support in the second half of the eighteenth century among philosophers such as Cornelius De Pauw (1739–1799) and Henry Home, Lord Kames (1696–1782). Home even posited the existence of distinct species within humankind—a notion strongly opposed by both Buffon and Blumenbach (De Angelis, 2024). However, *dégénération* cannot

[3] The Plain of Shinar is the name given to the southern region of Mesopotamia according to the Hebrew Bible, where Nimrod is said to have established a monarchical government (Goguet 1759, Part 1, Book 1, 24–25).

entirely be equated to "savagery" or "barbarism," as humanity's obscured knowledge could still be expanded and perfected. In fact, Goguet proposes core elements of the so-called stadial theory of the advancement of human societies. For him, this development consisted primarily of two stages: hunter-gatherer societies and agricultural societies. Thus, the central subject of his work concerns modes of subsistence, that is, the ways in which humans ensure survival through techniques and practices including hunting, fishing, agriculture, and the domestication of wild animals. In Goguet's work, stadial theory draws, once again, on the theory of human knowledge. He argues that the precious seeds ("*ces germes précieux*") of the most useful and necessary knowledge were preserved among the early settlers in the Plain of Shinar who subsequently migrated to Persia, Syria, and Egypt, where they established the first great civilisations (Goguet, 1759, Introduction, 10–11). This is also why Goguet identifies these societies as the starting point for the advancement of civilisation rather than, as was otherwise typical in the eighteenth century, ancient Greece. He considers artisan techniques ("*les Arts*") the product of sedentary societies and agriculture (Goguet, 1759, Preface, XXV). With his approach to the origins of laws, artisan techniques, and sciences, Goguet thus offers a counter program to Rousseau's hypothesis on the state of nature. In *Discours sur l'Origine et les Fondemens de l'Inégalité parmi les Hommes* (1755), Rousseau spoke of eliminating all (biblical) facts and based his own views on the methods of hypothesis formation as employed by natural scientists—e. g., on Buffon's theory of the Earth's formation in the first volume of *Histoire naturelle* (1749)—discussing facts of natural history primarily in footnotes (Rousseau, 2008, Exordium, 71–72).[4] In contrast, Goguet bases his arguments on the order of historical facts ("*l'histoire & l'ordre des faits*"), stating that otherwise, one would (as with Rousseau) follow one's own ideas rather than actual events, subordinating reality to imagination (Goguet, 1759, Preface, XXX–XXXI; Wolloch, 2007). It is important to note that Goguet's cultural-historical method was still present around 1790. As I will show, Blumenbach's characterisation of the (physical) *dégénération* of non-European peoples is also connected to the theory of stadial theory. In this context, he references both the material culture and the cultural systems of the so-called savage society ("les mœurs des peuples sauvages").

2.3 The Naturalisation of History and the "Ignoble-Savage" Thesis

Stadial theory as a theory of progress was from the beginning linked to natural law theory, as for instance in Samuel Pufendorf's (1632–1694) work (Hont, 2005). It subsequently played a central role in the development of Adam Smith's (1723–1790) economic theory (touching on ideas of property, the division of labour, commercial society, and self-interest, etc.) (De Angelis, 2021). It is no coincidence that Goguet

[4] One particularly well-known passage is note X on the varieties of humankind and the so-called "wild people" in the state of nature, who supposedly did not speak.

starts Book 1 of *De l'Origine des Loix, des Arts, et des Sciences* with a discussion of natural law ("*loi naturelle*"), which he describes as innate moral judgment embedded in the hearts of all people ("*ces sentimens de justice & d'équité*") (Goguet, 1759, Part 1, Book 1, 15). Once again, it becomes apparent that, while his general framework remains biblical, Goguet's source-based considerations address human progress. In his chapter on agriculture, Goguet provides concrete examples of how, from the earliest centuries onwards, families in Mesopotamia, Palestine, Egypt, Greece, and later in the southern provinces of France, worked the land with ploughs to ease their physical labour ("*labourage*"). He also describes how they used fertilisers ("*fertilisation*") and developed bread-making techniques ("*l'art de faire le pain*") (Goguet, 1759, Part 1, Book 2, Chapter 1, 180–222).

The reception of Goguet's *De l'Origine des Loix, des Arts et des Sciences* in Scotland (where it was available in English by 1761) was significant for the exploration of models of history among Scottish philosophers (Sebastiani, 2013, 79). However, their interpretations of natural law diverged: Adam Ferguson (1723–1816) draws on Montesquieu's concept of the "*esprit général des nations*" from *De l'esprit des lois* (1748). In his *Essay on the History of Civil Society* (1767), he renders the universality of abstract natural law according to the dynamic "*esprit général*" of civil society, attributing the mode of subsistence, and thus a peoples' ways of life, primarily to geography and climate (Starobinski, 2024, 129–130; De Angelis, 2024, 41–42). In doing so, Ferguson translates the historical concept of "progress" into biological terms, asserting that "in the material world, subjects organized as progressive [...] are characterized by the enumeration of steps in the passage from one form or state of existence to another, [...] the natural state of a living creature includes all its known variations, [...] through which life in its varieties is known to pass" (Ferguson, 1792, 190–192).

Prompted by Buffon's *Histoire naturelle* (1749), towards which he adopts a historical-philosophical perspective, Ferguson views the goal of the civilisation process as open-ended and fundamentally not subject to rational planning (Ferguson, 1768, 187; De Angelis, 2022, 63–64). In stark contrast, Henry Home, Lord Kames (1696–1782), introduced a novel perspective into the progress debate in his *Sketches of the History of Man* (1778, 2nd edition). Home argued that the history of humanity diverged due to the varied "natures" of peoples (Sebastiani, 2013, 78), which he believed he could classify as different species of humankind (Home, Lord Kames, 1778, 2nd edition).[5] As Goguet suggested, stadial theory had a dehumanising potential, which in the 1770s was explicitly activated with regard to Indigenous peoples of the Americas. This interpretative approach, essentially based on the "ignoble-savage" thesis (Meek, 1976), was explored from different perspectives in Home's *Sketches*, William Robertson's *History of America* (1777), and Cornelius De Pauw's *Recherches Philosophiques sur les Américains* (1769–1771). By way of correcting the Bible, Home suggested "separate races" and local acts of creation, Robertson considered the entire American continent as populated by a uniform human type,

[5] Home (Lord Kames) (^{2}1778), B. 1, 20: "If the only rule afforded by nature for classing animals can be depended upon, there are different species of men as well as of dogs [...]."

and De Pauw deemed the Americans to be particularly "degenerated" (Sebastiani, 2013, 87–95). In this context, De Pauw still used the term '*verartet*' ('degenerated') according to the theory of *dégénération* as put forward by Buffon (De Angelis, 2024, 43), whose concept of species Home also sharply polemised (Sebastiani, 2013, 81). The key point in this interpretation of stadial theory is linked to a systematic set of prejudices that had been invoked against Indigenous American peoples, targeting aspects of their origins, perceived stagnation, supposed stupidity, moral characteristics, sexual impotence, weakness, infertility, apathy, insensitivity, and more. Blumenbach countered these prejudices in *De generis humani varietate nativa* (1795) and attempted to expose them as unfounded. He, instead, regarded the *Varietas Americana* as a "savage" society with its own customs, rituals, and rules. Blumenbach concluded that all varieties of humankind are likely parts of one single species (Blumenbach, 1795, § 90). Thus, his *Conclusio* should also be understood against the backdrop of the "ignoble-savage" thesis. Even more importantly, this perspective moves beyond simply restoring the monogenesis idea, which is often noted as being more aligned with the biblical account. Instead, it points to a new conception of natural history that emerged in the 1780s and 1790s and in which the Blumenbach-Kant debate on the idea of the formative drive plays a central role.

2.4 The Notion of the Formative Drive and a New Archaeology of the Globe

The relationship between Blumenbach and Kant can essentially be described in three phases, each marking significant developments in their conceptualisation of natural history. The first phase involves Kant's systematic reflections on the notion of the "ladder of continuity among creatures," in which he incorporates Leibniz's concept of nature (*natura non facit saltus* [nature makes no leaps]) into the principle of the continuity of forms. This principle unites the two other principles of reason—homogeneity and specification—and completes the systematic connection, both in the ascent to higher genera and in the descent to lower species, through an overarching idea (Kant, 1998, 598, AA III 435 = A657/B685). The ladder of continuity among creatures thus constitutes "a system, that is, a whole of cognition ordered according to principles" (Kant, 2002, 183, AA IV, 467). For Kant, however, "The rungs of such a ladder [...] stand too far apart from one another, and what we presume to be small differences are commonly such wide gaps in nature itself" that we cannot deduce "the intentions of nature" from experience (Kant, 1998, 604, AA III, 441 = A668 / B696). The ladder of continuity among creatures, nevertheless, remains a "regulative principle of reason" that clearly transcends experience and provides a unified systematic framework for the empirical investigation of gaps in nature (Kant, 1998, 604, AA III, 442 = B696). The second phase in the relationship between Blumenbach and Kant is marked by Kant's reference to Blumenbach's natural history. In his *Handbuch der Naturgeschichte* (*Handbook of Natural History*) (1779), Blumenbach called into question "the familiar images of chains and ladders and nets which have been applied to nature" (Part 1, § 7, 10, my translation). For Blumenbach, "the idea of the ladder of continuity in nature" remains speculative and leads to imaginary classifications of realms, categories, and classes of living beings. He considers it "the true impetus for a natural system in natural history," which naturalists desire and according to which one could classify natural bodies by degrees of kinship (§ 7, 11, my translation). Moreover, he identifies the functionality of living beings' body structures and their integration into the food chain—from the earth to plants, animals, and humans, and back—as the reason why "creation follows its course" (§ 7, 13, my translation); "within this great cycle, reason does not need to assume links [among the three kingdoms of nature]" which experience itself has thus far also failed to reveal (§ 7, 14, my translation). In other words, the cycle has replaced the ladder of continuity; and even from a purely geometric perspective, questions regarding origin and transitions make little sense within the context of a circle. With reference to Blumenbach, Kant had thus distanced himself from the idea of the "chain of nature" (Kant, 2007, 214, AA VIII, 180), while, at the same time, Kant highlights another one of Blumenbach's central ideas or concepts: the "formative drive" (*Bildungstrieb*) (Kant, 2007, 214, AA VIII, 180). The formative drive takes effect directly on the level of organised matter, while questions pertaining to the origin of and transitions from inorganic matter (atoms, molecules) to the organic structures of living beings remain (initially) unaddressed. Blumenbach thus implies a leap in the ladder of continuity by recognising a gap between organised and inorganic nature, which fundamentally disavows the ladder of continuity. The third phase of the Kant-Blumenbach relationship is, finally, characterised by the transformation of natural history into an archaeology of nature. This shift originates from Blumenbach's fossil studies, which observe the extinction of prehistoric organisms as well as the formation of *altered* organisms in the modern world by virtue of the formative drive. From this shift, Blumenbach derives the thesis of nature's mutability (Blumenbach, 1790, 1790a). More or less simultaneously, Kant introduces the rational principle of teleological judgment, which lends a unified structure to the study of the causes and laws governing the formation of living beings (Kant, 2000, 285–286 = AA V, 416–417). In this case, teleology thus serves as a regulative idea. Accordingly, Sects. 81 (Formative Drive and Epigenesis) and 82 (Geohistory) in the *Critique of the Power of Judgment* form a systematic connection. That is to say that Blumenbach's model of Earth's revolutions serves as the premise for Kant's synthesis of mechanical and teleological modes of explanation, which does not determine the laws of formative nature as such but merely permits considerations along the lines of the regulative principle of final causes (Kant, 2000, 296–297; De Angelis forthcoming). The question regarding the origin and transformation of living beings thus shifts towards the *temporal* dimension of Earth's history.

The reflections by Kant and Blumenbach largely align with John Pocock's observations on the characteristics of eighteenth-century theory building processes: the eighteenth century is neither aware of the science of geology, nor does it discover human fossils. Nevertheless, this does not prevent naturalists from proposing geohistorical hypotheses and supporting them with geological evidence. Beyond his universalistic thesis of total revolution and his 1790 fossil studies on encrinites (De Angelis forthcoming), Blumenbach penned an essay on the archaeology of the globe in 1803 (Blumenbach, 1803; Rudwick, 2005, 424–428), which attracted interest among French naturalists. For instance, Antoine-Marie Héron de Villefosse (1774–1852), a mining engineer and commissioner in the Harz during the Napoleonic occupation, published a detailed extract of Blumenbach's essay in the *Journal des Mines* (de Villefosse, 1803). Blumenbach classified fossils according to a chronological method, associating various classes of fossils with different revolutions of the globe: while the first and second classes belong to a more recent period in Earth's history and can be compared with living specimens, the third class includes so-called "dubious" fossils, such as bones of large quadrupeds (rhinoceros). With the third class, it remains uncertain whether the respective fossils and the organised bodies resembling them can be traced back to the same species that may simply have undergone a slight degeneration (*dégénération*), or if they belong to a different species altogether (Blumenbach, 1803, 12; de Villefosse, 1803, 11). Finally, Blumenbach counts *phytolithes*—i.e., fossils found in the greywacke slate of the Hanover region—as the fourth class, i.e. the pre-Adamic class of fossils, belonging to creatures which counted among the earliest traces of organisms on our planet (Blumenbach, 1803, 20; de Villefosse, 1803, 27). Héron de Villefosse highlights the following innovative idea Blumenbach put forward in his conclusion: Blumenbach observed that the reproduction of organised bodies, in physiological terms, relates to the environment in various ways ("*nisus eorum formativi multifaria esse relationem*") as regards material constitution, diet, climate, etc. (Blumenbach, 1803, 26; de Villefosse, 1803, 34). Among organised bodies, some are therefore changeable, adapt readily to altered conditions of their existence, and eventually degenerate (*dégénération*), while others are more constant in their form and existence and maintain themselves in almost any climate (Blumenbach, 1803, 26–27; de Villefosse, 1803, 34–35). Thus, Blumenbach applies the concept of *dégénération* as a neutral term synonymous with changes in earth-historical processes. For Blumenbach, nature itself hence changed everything ("*natura quae omnia commutat et vertere cogit*") by way of adapting to the major catastrophes having affected the planet. It preserved some animals and plants from the first creation on the restored Earth, while replacing the majority of extinct individuals with other genera and species ("*novis generibus et speciebus*") better suited to the globe's new conditions (Blumenbach, 1803, 27; de Villefosse, 1803, 35). Blumenbach concluded his essay with a quote from Lucretius's *De rerum natura*: "[M]undi naturam totius aetas/ Mutat/ et ex alio terram status excipit alter./ Quoad potuit, nequeat; possit quod non tulit ante." ["The nature of the whole world changes with time; one state of the earth follows another. What once was possible, now is not; what was once impossible, now is [possible]."] (Blumenbach, 1803, 27; de Villefosse, 1803, 35, my translation). This Lucretius quote is neither to be understood in atheistic terms nor does Blumenbach advocate a randomness theory of the cosmos. Instead, Blumenbach upholds a transformist view of nature, which is rooted in the category of time. This view also characterises Blumenbach's concept of natural history, which presupposes *dégénération* as the alteration of organic bodies and includes an explanation of the varieties within humankind.

2.5 *Dégénération* and the Formative Drive

In his treatise on the varieties of humankind, Blumenbach dedicates a central chapter to the concept of the formative drive, drawing on Kant's *Critique of the Power of Judgment* (§ 82), whose argument he essentially adopts. From a methodological perspective, Kant had discussed the possibility that reason can conceive both mechanical and teleological explanations for the development of organic bodies, *quasi* as a subjective condition of our use of reason (Blumenbach, 1795, § 33, 82–83). There is an explicit reference to Blumenbach's treatise *Über den Bildungstrieb* (*On the Formative Drive*) (1789, 2nd edition; 1791, 3rd edition), which Kant had received favourably. Blumenbach elaborates how the formative drive can deviate from its intended direction and norm (*directione et norma*) in three ways: first, by bringing forth monstrosities (*monstruosa fabrica*); second, by generating hybrids from interbreeding reproductive material of different species (*hybrida generatio*); and third, by degeneration into so-called varieties (*degeneratione in varietates*) (Blumenbach, 1795, § 33, 85). Although hybrids, by a natural law, are not typically capable of reproduction, for Blumenbach they, nonetheless, serve to refute the theory of preformation of germs and demonstrate the effects of the formative drive. To substantiate this argument, Blumenbach draws on botanist Joseph Gottlieb Kölreuter's (1733–1806) experimental studies on the gender of plants. In the rare cases of reproducing plant hybrids, they were fertilised with the male pollen of the same species over several generations, with multiply repeated fertilisation. In the great-grandchild generation, the permutation showed a deviation from the original maternal form (*materna forma*) and a gradual shift toward the paternal form (*paternam alterius speciei formam*), such that the former seemed to have transitioned into the latter as if by arbitrary metamorphosis (Blumenbach, 1795, § 33, 86–87).[6] Blumenbach regarded this result as a refutation of the theory of preformation in its ovist variant, which held that germs are preserved in the mother's ovaries. He interpreted the transformation of the plant as an effect of the formative drive, prompted to change its course by an external human intervention. This plant example thus illustrates Blumenbach's stimulus–response model of organic bodies, which he considered the cause of *dégénération* (Blumenbach, 1795, § 32, 81–82). On a larger spatiotemporal scale, the permanent, prolonged, and continuous influence of stimuli on organic bodies, sustained through a long series of generations, also contributes to gradually diverting the formative drive from its usual course. In this, Blumenbach saw the most common cause of *dégénération*, as well as the actual source of the varieties of humankind (Blumenbach, 1795, § 33, 87–88). These external stimuli are, notably, climate, food, and the modes of life displayed by humans as they spread across the various (known) continents of the Earth over the course of millennia. Thus, Blumenbach synthesises Goguet's migration theory and the physiological theory of *dégénération* triggered by the formative drive.

[6] See also Blumenbach 1795, § 33, fn. *G*, my translation): "Kölreuter's *third continuation of news on experiments concerning the gender of plants* & c. p. 51, § 24: 'Complete transformation of one natural plant genus into another.'".

2.6 Geographical History and the Comparison of Human Faces and Skulls

The topic of facial structures provides an opportunity to examine Blumenbach's scientific method, which illustrates how he developed his scholarship and presented it in his writings. Blumenbach distinguishes five facial varieties (*varietates faciei gentiliciae*), with the term "gentilicium" referring to *gens*, *stirps*, i.e., to a people, lineage, or family—a denotation already common in ancient Rome: (i) the oval, European variety; (ii) the Asian one ("the Tartar face"); (iii) the American ("*facies Americanorum*"); (iv) the African ("the Guinea face"); and (v) the Malay ("*Malaica stirpis facies*") (Blumenbach, 1795 § 56, 176–182). In the eighteenth century, the classification of facial shapes was an epistemic practice similar to the classification of minerals, plants, and animals. Blumenbach, moreover, introduces an aesthetic evaluation by describing the European facial form as "the most beautiful and well-proportioned one according to our judgment of symmetry" ("*ex nostratim de symmetria judicio*") (Blumenbach, 1795, § 56, 178). This judgment was, hence, relative to European standards of beauty which, in the eighteenth century, were still influenced by the ancient Greek concept of beauty in human nature, rooted in classical art and theoretically linked to climate theory (Winckelmann, 2002, 2nd edition, 244–249). Blumenbach cites climate as the primary cause of different facial forms, with the temperate zone from the Ural Mountains to the Atlantic playing a central role in the development of European features. At the same time, Blumenbach points to examples of people who, over time ("*temporis progressu*") and through settling in new climates, changed their facial features. Historians, for example, considered the Yakuts as a branch of the Tartars. However, observant eyewitnesses reported ("*curati autoptae asserunt*") them to have Mongolian facial features, which Blumenbach himself observes ("*coram video*") in the skull of a Yakut he received from Freiherr B. von Asch (1729–1807) and which he included in his anthropological collection (Blumenbach, 1795, § 57, 185). Creoles, similarly, originated as English settlers mixed with Native Americans, and the characteristic facial features of the Indians who once conquered the Egyptian peninsula can only be recognised in statues unearthed from subterranean temples housed in the British Museum among the ancient treasures of antiquarian and collector Charles Townley (1737–1805) and which Blumenbach was able to view (Blumenbach, 1795, § 57, 185–187; Gascoigne, 2010).[7] Blumenbach, moreover, notes that the Mongols who conquered India, from Timur's time onwards, gradually lost their original facial traits and took on Indian features, a change documented by naturalist and secretary to the Governor of Bengal, John Walsh (1726–1795), who provided Blumenbach with visual evidence ("imaginibus Indicis ad oculum mihi demonstravit") (Blumenbach, 1795, § 57, 187). Blumenbach's knowledge is thus

[7] "When Blumenbach visited London in 1791 it was Banks who enabled him to view specimens in private collections as well as the British Museum. The link between such collecting and the English gentlemanly virtuoso tradition is apparent in Blumenbach's praise to Banks of the Egyptian artefacts in the collection of 'your great Antiquarians Mr. Townley & Mr. Knight' BL, Add 8098, fos. 213–4, Blumenbach to Banks, 8. Jan. 1794)'" (Gascoigne 2010, 91).

primarily composed of autopsy, testimonies, artworks, private collections, museums, skull specimens, the theory of migration, and an aesthetic judgment. The aesthetic judgment concerning the beautiful facial structure of Europeans, however, does not originate from Blumenbach himself, or at least, cannot be attributed to him alone. Blumenbach actually references the French explorer of the Near East, Jean Chardin (1643–1713), who, following the model of eyewitness observation, wrote the following in his *Journal Du Voyage* (1686):

"The complexion of the Georgians is the most beautiful in all the earth; and I can safely say that I never saw an ill-favoured countenance in all that country, either of one or other sex; but I have seen those that had Angels Faces; nature having bestowed upon the women of that country graces and features which are not other where to be seen, So that it is impossible to behold them without falling in love. More charming countenances, nor more lovely statures and proportions cannot be pencilled forth by all the art of man than those of the Georgian women." (Chardin, 1686b, 190; Blumenbach, 1795, § 85, 303, Fn. *s*).[8]

The maximal symmetry and beauty of the skull of a Georgian woman is thus defined as the most beautiful European facial form. This definition arises from the discussion on measuring cranial variations, in which Blumenbach as anatomist introduces his "vertical norm" (*norma verticalis*), thus setting himself apart from Pieter Camper's facial angle measurement (*linea facialis*) (Blumenbach, 1795, § 61, 203–205). Blumenbach famously illustrates this paragraph with the well-known image of the five main human varieties' respective skull forms on a horizontal line, with the skull of a Georgian woman at the centre (cf., i.a., Schaub, Sebastiani, 2021, 429–433, Junker, 2019; Böker, 2019; Doron, 2016, 386–387). The Georgian woman's skull thus represents the midpoint between the extreme poles, i.e., the cranial forms of Mongolian and African peoples (Blumenbach, 1795, § 62, 206–207). With regard to the question of why Blumenbach argues and classifies in this way, the following factors must be considered. He categorises the human race into five primary varieties: (A) Caucasian, (B) Mongolian, (C) Ethiopian (African), (D) American, and (E) Malay; for reasons to be discussed, he ranks the Caucasian variety, considered the original variety, first (Blumenbach, 1795, § 81, 285–286). Once again, Blumenbach invokes the notion of the Caucasian variety as—by European standards—the most beautiful facial form. This variety includes Europeans (except the Lapps and Finns), Western Asians ("*Asiani occidentales*") settled from the River Ob via the Caspian Sea to the Ganges, as well as the inhabitants of northern Africa (Blumenbach, 1795, § 82, 289–290). Blumenbach points out that other authors have likewise divided humankind into varieties, mentioning, e.g., Leibniz, Linnaeus, Buffon, Governor Pownall, Abbé de la Croix, Kant, John Hunter, E. A. W. Zimmermann, Chr. Meiners, G. S. Klügel, and J. D. Metzger. His aim in doing so is to enable the reader to compare, assess, appreciate, and choose among the various perspectives (Blumenbach, 1795, § 83,

[8] "Le sang de Georgie est le plus-beau d'Orient, & je puis dire du monde, je n'ai pas remarqué un laid visage en ce pais-la, parmi l'un & l'autre sexe: mais j'y ai vû d'Angeliques. La nature y a répandu sur la plus-part des femmes des graces, qu'on ne voit point ailleurs. Je tiens pour impossible de les regarder sans les aimer. L'on ne peut peindre de plus-charmans visages, ni-de plus belles tailles qu'on les *Géorgiennes*." (Chardin 1686a, 238–239).

296–302). Publishing research results, series of experiments, research protocols, etc., to enable readers to form their own judgments on controversial issues had been an established approach since the "scientific revolution," practiced by figures such as Robert Boyle, Charles Bonnet, and Albrecht von Haller. This approach was based on an epistemology of credibility and trust and was part of the theory of authority and testimony which Blumenbach observed alongside autopsy (De Angelis, 2013, 2024). One reason Blumenbach considered the Caucasian variety as the original may stem from the research approach of physicist and geographer Eberhard August Wilhelm Zimmermann (1743–1815), whose *Geographische Geschichte des Menschen und der allgemein verbreiteten vierfüßigen Thiere* (1778–1783) was a crucial publication for late eighteenth-century natural history, which has, however, received relatively little attention so far. Thus, Blumenbach employs both an aesthetic as well as a geographical argument: Zimmermann posits the origin of the human race in the Caucasus, where "one of the greatest habitable elevations of our Earth" is located and from where "the largest mountain ranges expand", making it "quite natural for the varieties to derive from here" (Zimmermann, 1778, 114; Blumenbach, 1795, § 83, 300, fn. *o*, my translation). The key point is that, regarding the geographical diffusion of the varieties, Zimmermann's theory aligns with Goguet's migration theory and with the concept of degeneration as per climate theory, while also further supporting them with scientific data. An increase in population compelled people to seek larger habitable areas, leading families to spread across different regions of the world—from the Ural to the Caucasian Mountains, into Europe, along the northern edge of the Asian continent as far as Siberia, to North America, Arabia, India, and the associated islands (perhaps Africa was then connected to Asia by a large landmass).

"Human beings gradually adapted to the climate; they became Negroes. However, this could have also started from Europe. Eventually, a southeastern colony gave rise to the Chinese, Koreans, and the local populations. Thus, as humanity was gradually transferred to all climates, the climatic influences transformed them—here into Georgians, there into Negroes, and elsewhere into Eskimos." (Zimmermann, 1778, 115, my translation).

Migration theory would, furthermore, explain why the Caucasians feature white skin and why it is considered the original "colour" of the human race. According to Blumenbach, as families migrated to different regions, the white colour darkened more easily than it lightened ("*facilis in fuscum degeneratio sit*") (Blumenbach, 1795, § 85, 304). Finally, it is notable that Zimmermann, like Goguet, does not primarily focus on individual men or women but rather on families, who, through reproduction and adaptation to new climate zones, ensured the continued existence of the human species. The connection to Zimmermann's *Geographische Geschichte des Menschen* shows that Blumenbach's model of classifying human varieties was embedded in a broader framework of considerations by naturalists who (in contrast to Rousseau and other philosophers like Kant or Meiners) aimed to support a new model of human history, also by substantiating it with natural historical evidence.

2.6.1 Varietas Americana

Varietas Americana (the "American Race") is a central chapter in Blumenbach's treatise on the varieties of the humankind. It further evolves Goguet's cultural-historical approach, expanding, however, on the classical authors with a wealth of ethnographic data on non-European peoples put forward in the latter half of the eighteenth century by historians, philosophers, travellers, sailors, and natural scientists. At the same time, the chapter on the *Varietas Americana* evidences the methodological principles guiding Blumenbach's critical argumentation, primarily aimed at dispelling prejudices about Native Americans. Blumenbach appears quite astonished at the sheer number and nature of figments ("*quanta et qualia figmenta*") about the Native American variety that circulate among the writings of philosophers, historians, and travellers, concerning, for instance, the beardlessness (Cornélis de Pauw), menstruation (Martin Schurig, 1729), skin colour (Henry Home), or facial features (William Robertson) of Americans (Blumenbach, 1795, § 88, 311). His first criterion against such biases includes observation, experience, autopsy, and credible eyewitness testimony as per the doctrine of authority and testimony. The second criterion is the cultural system of Native Americans, which had already played a central role in José de Acosta's (1540–1600) *Historia Natural y Moral de las Indias* (1590): "The category of system meant that within the different areas of social life, any fact (the methods of dressing, preaching, eating, etc.) was a social fact that was linked to all others, so that these various areas—religion, ethics, politics—were mutually connected" (Imbruglia, 2014, 303). Acosta also defined "savage society" as a form of civil society as yet unorganised in terms of state structures but nevertheless with "social rules" that elevate it above a state of "barbarism" (Imbruglia, 2014, 303). Both Goguet's and Blumenbach's texts consistently relate to Acosta's *Historia Natural*. This second criterion of a cultural system, finally, leads to a third one: the time or duration by which a *sauvage society* evolves according to the theory of progress. I next reconstruct how these three criteria interrelate in Blumenbach's work and how the factor of time plays a central role in the development of the *Varietas Americana*.

To refute the claim that Americans are beardless, Blumenbach cites witnesses who confirm that, across the Americas, from the Eskimo regions to the Tierra del Fuego, there are entire tribes who grow beards, citing Johann Georg Gmelin's *Reise durch Sibirien* (1733–1743): "It is rare to find a beard among the Tungus and all these peoples. For as soon as it appears, they pluck out the hairs and eventually achieve that none grow back" (Blumenbach, 1795, § 88, 312, fn. *I*, my translation; Gmelin, 1751). Furthermore, he exposes the myth of "menstrual absence" among American women, which arose as Europeans had allegedly never witnessed such a process (Blumenbach, 1795, § 88, 313, fn. *l*; de Léry, 1578). Blumenbach, however, cites Gabriel Sagard's *Voyage du pays des Hurons* (1632), which states that menstruating women were considered poisonous ("*pro venenatis quasi habentur*") and thus removed to remote huts to enjoy rest away from the male gaze (Blumenbach, 1795, § 88, 314, fn. *m;* Sagard, 1632). Blumenbach further ascribes the wider than merely black variations in skin colour among Americans to the influence of the climate ("*ex climatis Americani ratione*") and cites Zimmermann's *Geographische Geschichte des Menschen* (1778) as a source that attests to a white complexion among the inhabitants of the Andean region (Blumenbach, 1795, § 88, 314–315, fn. *o*; De Angelis, 2024, 12–13). However, Blumenbach also attributes American colour variations to the laws of *dégénération* ("*ex degenerationis legibus*"), proposing a likely origin of Americans in northern Asia, citing an article by Kant in the 1788 *Der Teutsche Mercur* (Blumenbach, 1795, § 88, 315, fn. *p*; Kant, 2007). Blumenbach, finally, also explores a similar approach regarding the facial structure of Americans ("*de vultu Americanos*"). He counters claims of homogeneity in the features of the New World's inhabitants by citing the eyewitness Abbé Giovanni Ignazio Molina, who notes in *Sulla storia naturale del Chili* (1782) that individuals within one nation ("*gl'individui di una nazione*") vary significantly from those of another: "[A] Chilean differs in appearance from a Peruvian no less than an Italian differs from a German" (Blumenbach, 1795, § 88, 315–316, fn. *q*; Molina, 1782, 336). Blumenbach also includes Nicolás del Techo (1654, 34, Caaiguos), Martin Dobrizhoffer (1783, Abipones), Antonio de Ulloa (1751, 204, Peruvians), and Georg Forster (1983, 920, Tierra del Fuego) among further credible eyewitnesses ("*tum autoptarum fide dignissimorum testimonia*") who observed the diversity of American facial features ("*vultus varietatem*") (Blumenbach, 1795, § 88, 316, fn. *r*). Finally, Blumenbach makes an important observation regarding the resemblance between Americans and Mongolians, a similarity noted as early on among the first European explorers, such as Amerigo Vespucci (1451–1512) (Blumenbach, 1795, § 88, 317, fn. *s*; Vespucci, 1745). The fluid boundaries and gradual transition between human varieties, distinguishable only by fine nuances, is indeed a central thesis in Blumenbach's skull studies. He considers this thesis confirmed by the theory of migration in connection with his own theory of Earth's revolutions and, thus, the factor of time: He argues that the Americans, originating from Mongolians, likely migrated from northern Asia in multiple waves interrupted by long intervals, possibly spurred by physical, geological, and political catastrophes (Blumenbach, 1795, § 88, 317). Furthermore, Blumenbach conjectures that distinctive Mongolian features are more pronounced among Eskimos than other Americans, firstly, because, driven by new catastrophes, they arrived later from northern Asia and, secondly, they now live in climates similar to their homeland (Blumenbach, 1795, § 18, 317–318). As evidence supporting this observation, Blumenbach cites two Eskimo skulls from his collection that originated from the colony of Nain in Labrador. Furthermore, he cites images of these "savages" ("*in imaginis horum barbarorum*"), painted from life ("*ad vivum*") by two distinguished artists and gifted to him by Joseph Banks, then president of the Royal Society (Blumenbach, 1795, § 88, 318, Fn. *t*). Since the Renaissance, images have served as means of illustrating claims to knowledge in line with the doctrine of authority and testimony (De Angelis, 2011). Moreover, regarding the thesis on the influence of climate on the facial features of various groups of people, Blumenbach mentions the mariner Jan Huygen van Linschoten (1563–1611) who had seen the inhabitants of the Magellan Strait firsthand and compared their physiognomy, facial features, skin colour, and hair with the Samoyed people of Siberia, whom he had encountered during his journey along the Nassau Route (Blumenbach, 1795, § 88, 318, Fn. *w*). I have illustrated how Blumenbach presents his claims about non-European peoples in *De generis humani varietate nativa*, and especially how he reasons for the temporalisation of the somatic traits of American populations. That he describes some of these peoples as "barbarians" reflects the fact that they were apparently still nomads and not sedentary, subsisting on fishing and hunting reindeer herds, as migration theory would posit.

2.7 Conclusion

This chapter has demonstrated that examining Blumenbach's natural history, and especially his writings on the varieties of humankind, requires thorough and careful historicisation, and must draw attention to the distinctions between the treatise's genesis and its nineteenth- and twentieth-century reception. Central to Blumenbach's argument is his critique of preformationism, against which he argues for nature's mutability, empirically supported by fossil research in southern Germany and around Hanover. His renowned symmetrical (and non-hierarchical!) classification scheme for human varieties resonates with the tradition of Carl Linnaeus. Moreover, Blumenbach's explanations of human varieties with regard to climate, heredity, and the factor of time align with the programme of the progressive late Enlightenment natural history. Blumenbach's adherence to the unity and equality of humankind is, finally, in line with a central Enlightenment concept: natural law theory, following the tradition of Samuel Pufendorf, with sociability as its core tenet (De Angelis, 2024). That is to say, Blumenbach expressed views not only about physical differences but also about the mental capacities of specific human varieties, especially concerning Africans who had escaped their fate as slaves and received education and training in England. Blumenbach sharply criticised the slave trade, including "the bestial brutality of the [...] white executioners on the slave ship[s] and in the West Indian sugar plantations" (Blumenbach, 2014 [1790], 91). True to the principles of authority and testimony, he allowed the actual voices of African writers, poets, and philosophers—such as Anton Wilhelm Amo, Ignatius Sancho, and Gustav Vassa (in contemporary historical texts known as Olaudah Equiano)—to be heard directly (Blumenbach, 2014 [1790], 85–118). Blumenbach chose them according to criteria of perfectibility and cultural capacity, as he points out (not without irony) towards some of his contemporaries: "I do not know any *savage* people who *as much as the Negroes* distinguished themselves by such examples of perfectibility and even scientific cultural capacity that they align themselves so closely with the most cultivated nations on earth" (Blumenbach, 2014 [1790], 118, my translation). Thus, Gustav Vassa (d. 1797), for instance, regarded slavery as a state of perpetual "warfare"; and Atlantic slavery became part of broader geopolitical networks, conflicts, and spheres of interest when the Jamaican slave revolt of 1760–1761 was recognised as warfare of sorts (Brown, 2020, 3–9). Historicising Blumenbach's treatise on the varieties of humankind also means to acknowledge this geopolitical dimension as part of his natural history. Our current debate regarding *race* in the eighteenth century almost entirely neglects this very aspect of natural law within the context varieties of humankind (De Angelis, 2024).

References

Acosta, J. de. (1590). *Historia Natural y Moral de las Indias*. Juan de Leon.

Blumenbach, J. F. (1803). *Specimen Archaeologiae Telluris Terrarumque Inprimis Hannoverarum*. Apud Henricum Dietrich.

Blumenbach, J. F. (1795). *De Generis Humani Varietate Nativa*. Editio Tertia. Apud Vandenhoek et Ruprecht.

Blumenbach, J. F. (1791). *Über den Bildungstrieb* (3rd ed). Bey Johann Christian Dieterich.

Blumenbach, J. F. 2014[1790]. *Beyträge zur Naturgeschichte* (M. Marino, Ed.). Olms-Weidmann.

Blumenbach, J. F. (1790a). Beyträge zur Naturgeschichte der Vorwelt. *Magazin für das Neueste aus der Physik und Naturgeschichte*, Sechster Band, Viertes Stück. 1–15.

Blumenbach, J. F. (1789). *Über den Bildungstrieb*. (2nd ed). Bey Johann Christian Dieterich, 1789.

Blumenbach, J. F. (1779). *Handbuch der Naturgeschichte*, Dritte sehr verbesserte Ausgabe. Bey Johann Christian Dieterich.

Böker, W. (2019). Blumenbach's collection of human skulls. In N. Rupke and G. Lauer (Eds.), *Johann Friedrich Blumenbach. Race and Natural History, 1750–1850* (80–95). Routledge.

Brown, V. (2020). *Tacky's Revolt. The Story of an Atlantic Slave War*. The Belknap Press of Harvard University Press.

Comte de Buffon, G. L. Le Clerc (1749). *Histoire Naturelle, Générale et Particulière, Avec la Description du Cabinet du Roi* (Vol. 1). De l'Imprimerie Royale.

Chardin, J. (1686a). *Journal Du Voyage Du Chevalier Chardin en Perse & aux Indes Orientales, par la Mer Noire & par la Colchide*. Qui contient Le Voyage de Paris à Isphahan. Chez Daniel Horthemels, Libraire, rue de la Harpe, au Mécénas.

Chardin, J. (1686b). *The travels of Sir John Chardin into Persia and the East Indies: The first volume, containing the author's voyage from Paris to Ispahan to which is added, the coronation of this present king of Persia, Solyman the Third*. Printed for Moses Pitt in Duke-Street Westminster.

Danneberg, L. (2009). Kontrafaktische Imaginationen in der Hermeneutik und in der Lehre des Testimoniums. In L. Danneberg et al. (Eds.), *Begriffe, Metaphern und Imaginationen in Philosophie und Wissenschaftsgeschichte* (pp. 287–449). Harrassowitz.

De Angelis, S. (forthcoming) "Johann Friedrich Blumenbach". In R. Meer and G. Stiening (Eds.), *Der Begriff der Idee von Kant bis Fries. Texte und Kommentare*. De Gruyter.

De Angelis, S. (2024). Das Menschenbild der 'Wissenschaft vom Menschen' (1660–1800). In M. Zichy (Ed.), *Handbuch Menschenbilder*, (pp. 33–49). Springer VS. https://doi.org/10.1007/978-3-658-32138-3_9-1

De Angelis, S. (2022). Natur, 'Kraft' und Geschichte. Kraftkonzeptionen in den 'Wissenschaften vom Leben' und ihre Bedeutung für die Analogie von Natur und Geschichte in der Spätaufklärung. In F. Fehrenbach et al. (Eds.), *Form- und Bewegungskräfte in Kunst, Literatur und Wissenschaft* (pp. 51–76). De Gruyter.

De Angelis, S. (2021). The concept of self-interest in eighteenth century anthropology and economic theory. From Richard Cumberland to Adam Smith. In C. Zabel (Ed.), *Historicizing Self-Interest in the Modern Atlantic World. A Plea for Ego?* (pp. 95–116). Routledge.

De Angelis, S. (2013). Experiments, Judicial Rhetoric and the *Testimonium*. Practices of Demonstration in the Hamberger-Haller Controversy on the Respiration Mechanism. In A. Holenstein et al. (Eds.), *Scholars in Action. The Practice of Knowledge and the Figure of the Savant in the 18th Century* (Vol. 2, pp. 679–701). Brill.

De Angelis, S. (2011). Sehen mit dem physischen und dem geistigen Auge. Formen des Wissens, Vertrauens und Zeigens in Texten der frühneuzeitlichen Medizin. In H. Jaumann (Ed.), *Diskurse der Gelehrtenkultur in der Frühen Neuzeit. Ein Handbuch* (pp. 211–253). De Gruyter.
De Angelis, S. (2004). Anthropologie und Gesetz. Konzepte von der Natur des Menschen im 16. Jahrhundert: Vives und Melanchthon. In B. Mahlmann-Bauer (Ed.), *Scientiae et Artes. Die Vermittlung alten und neuen Wissens in Literatur, Kunst und Musik* (Vol. 2, pp. 871–893). Harrassowitz.
De Léry, J. (1578). *Histoire d'un Voyage fait en la Terre Du Bresil, autrement dite Amerique.* Pour Antoine Chuppin.
Del Techo, N. (1654). Relatio De Caaiguarum Gente, Coepta ad Fidem Adduci. In *Relatio Triplex De Rebus Indicis* (pp. 32–47). Apud Iacobum Meursium.
De Pauw, C. (1769). *Recherches Philosophiques sur les Américains, ou Mémoires intéréssants pour server à l'Histoire de l'Espèce humaine* (Vol. 1). George Jacques Decker.
Dobritzhoffer, M. (1783). *Geschichte der Abiponer, einer berittenen und kriegerischen Nation in Paraguay.* Erster Theil. Bei Joseph Edlen von Kurzbek.
Doron, C.-O. (2016). *L'Homme Altéré. Races et Dégénérescence (XVIIe-XIXe Siècles).* Champ Vallon.
Ferguson, A. (1792). *Principles of Moral and Political Science* (Vol. 1). Printed for W. Strahan, and T. Cadell London, and for W. Creech.
Ferguson, A. (1768). *An Essay on the History of Civil Society.* The Second Edition. Printed for A. Millar and T. Cadell, in the Strand; and A. Kincaid and J. Bell.
Forster, G. (1983). *Reise um die Welt* (G. Steiner, Ed.). Insel.
Gascoigne, J. (2010). Blumenbach, Banks, and the Beginnings of Anthropology in Göttingen. In J. Gascoigne (Ed.), *Science, Philosophy and Religion in the Age of Enlightenment. British and Global Contexts* (pp. 86–98). Variorum.
Gmelin, J. G. (1751). *Reise durch Sibirien, von dem Jahr 1733 bis 1743.* Abram Vandenhoecks.
Goguet, A.-Y. (1759). *De l'Origine des Loix, des Arts, et des Sciences; et de leurs Progrès chez les Anciens Peuples. Depuis le Déluge jusqu'à la mort de Jacob.* Tome premier, Chez Desaint & Saillant.
Héron de Villefosse, A.-M. (1803). Considérations sur les Fossiles, Et particulièrement sur ceux que présente le pays de Hanovre. *Journal des Mines.* N° 91. Germinal An *12*, 5–36.
Home, H. (Lord Kames). (1778). *Sketches of the History of Man* (3 Vols) Printed for W. Strahan, and T. Cadell London, and for W. Creech.
Hont, I. (2005). The Language of Sociability and Commerce: Samuel Pufendorf and the Theoretical Foundations of the 'Four-Stages' Theory. In Id. *Jealousy of Trade. International Competition and the Nation-State in Historical Perspective* (pp. 159–184). Harvard University Press.
Imbruglia, G. (2014). "The Invention of Savage Society: Amerindian Religion and Society in Acosta's Anthropological Theology". *History of European Ideas, 40*(3), 291–311.
Junker, T. (2019). Blumenbach's theory of human races and the natural unity of humankind. In N. Rupke and G. Lauer (Eds.), *Johann Friedrich Blumenbach. Race and Natural History, 1750–1850*, (pp. 96–112). Routledge.
Kant, Immanuel. (2007). "On the use of philosophical principles in philosophy" (1788). In Id., *Anthropology, History and Education*, edited and translated by Robert B. Louden and Günter Zöller. (The Cambridge Edition of the Works of Immanuel Kant), edited by Paul Guyer and Allen W. Wood, 192–218. Cambridge University Press.
Kant, I. (2002). "Metaphysical Foundations of Natural Science" (1786). In Id. *Theoretical Philosophy after 1781*, edited by Henry Allison and Peter Heath, translated by Michael Friedman. (The Cambridge Edition of the Works of Immanuel Kant), edited by Paul Guyer and Allen W. Wood. Cambridge University Press.
Kant, I. (2000). *Critique of the Power of Judgement* (The Cambridge Edition of the Works of Immanuel Kant), edited and translated by Paul Guyer, translated by Eric Matthews. Cambridge University Press.

Kant, I. (1998). *Critique of Pure Reason* (The Cambridge Edition of the Works of Immanuel Kant) (P. Guyer & A. W. Wood, Eds.). Cambridge University Press.
Kant, I. (1900–). *Gesammelte Schriften.* In Preußische Akademie der Wissenschaften (Bd. XI, Briefe 1789–1794). Reimer.
Lafitau, J.-F. (1724). *Mœurs des Sauvages Américains Comparées Aux Mœurs des Premiers Temps* (Vol. 1). Chez Saugrain et Charles Etienne Hocherau.
de Maillet, B. (2023). *Telliamed. Entretiens d'un philosophe indien avec un missionaire français sur la diminution de la mer* (C. Cohen Ed.). Jérome Million.
Meek, R. L. (1976). *Social Science & The Ignoble Savage.* Cambridge University Press.
Molina, G. I. (1782). *Saggio Sulla Storia Naturale del Chili.* Stamperia di S. Tommaso d'Aquino.
Montesquieu, Charles-Louis, Baron de Secondat, et de La Brède et de Montesquieu. (1819). *Oeuvres* (Vol. 1) (Esprit des Lois, Vol. 1). Chez E.A. Lequien, Libraire.
Pocock, J. G. A. (2009). *Barbarism and Religion. Volume Four. Barbarians, Savages and Empires.* Cambridge University Press.
Robertson, W. (1780). *History of America* (Vol. 2). W. Strahan, T. Cadell.
Rousseau, J.-J. (2008). *Diskurs über die Ungleichheit/Discours sur l'inégalité* (H. Meier, Ed.). Ferdinand Schöningh.
Rudwick, M. (2005). *Bursting the Limits of Time. The Reconstruction of Geohistory in the Age of Revolution.* The University of Chicago Press.
Sagard, G. (1632). *Le Grand Voyage du Pays des Hurons, situé en l'Amérique vers le Mer douce, ès derniers confins de la nouvelle France, dite Canada.* Chez Denys Moraux.
Schaub, J-F., & Sebastiani, S. (2021). *Race et histoire dans les sociétés occidentales (XVe-XVIIIe siècle).* Albin Michel.
Schurig, M. (1729). *Parthenologia Historico-Medica, hoc est Verginitatis Consideratio, qua ad eam Pertinentes Pubertas & Menstruatio.* Apud Christophori Hekelii B. Filium.
Sebastiani, S. (2013). *The Scottish Enlightenment, Race, Gender, and the Limits of Progress.* Palgrave MacMillan.
Starobinski, J. (2024). *Montesquieu.* Seuil.
Ulloa, A. de, & Juan, G. (1751). Reise nach Südamerika. In *Allgemeine Historie der Reisen zu Wasser und Lande; oder Sammlung aller Reisebeschreibungen*, Bd. 9. Bey Arkstee und Merkus.
Winckelmann, J. J. (2002). *Geschichte der Kunst des Alterthums*, In A. H. Borbein et al. (Eds.), Philipp von Zabern.
Vespucci, A. (1745). *Vita e Lettere di Amerigo Vespucci* (A. Maria Bandini, Ed.). Nella Stamperia all' Insegna di Apollo.
Wolloch, N. (2007). "Facts, or Conjectures: Antoine-Yves Goguet's Historiography". *Journal of the History of Ideas, 68*(3), 429–449.
Zimmermann, E. A. W. (1778). *Geographische Geschichte des Menschen, und der allgemein verbre iteten vierfüßigen Thiere, nebst einer hieher gehörigen Zoologischen Weltkarte.* Erster Band. In der Weygandschen Buchhandlung.

Simone De Angelis is Professor of History of Science and Director of the Centre for History of Science at the University of Graz, Austria. His main research areas are Renaissance science and early modern history and philosophy of science and medicine, Enlightenment studies, history of life sciences, and anthropology. He has been Fellow of the Swiss National Science Foundation, the Herzog August Bibliothek in Wolfenbüttel, Visiting Researcher at the Wellcome Trust Centre for the History of Medicine at University College London and the Università degli studi di Padova. He is a Visiting Professor at École des Hautes Études en Sciences Sociales (EHESS) and the Centre Alexandre Koyré (CAK) in Paris. His books include Von Newton zu Haller (Niemeyer 2003), Scholarly Knowledge. Textbooks in early modern Europe (Droz 2008), Anthropologien (De Gruyter 2019, 2nd edition).

Chapter 3
Blumenbach's Conceptualization of Human Perfection and Its Evolution Through the History of Colonial Power

Victoria Shmidt

Abstract The ongoing evolution of Johann Friedrich Blumenbach's approach to human variation for the last two centuries can be interpreted through an understanding of race in science as inherently connected with the vicissitudes of the legitimacy of colonial power. In this chapter, a Blumenbachian vision on human perfection is historicized as reproduced in the various stages of colonial power. The circulation of Blumenbach's perspectives is connected to public and scientific debates regarding slavery and the post-slavery social order. The continuities and changes in Blumenbach's conceptualization of human perfection during post-slavery colonialism is elaborated through debates on racial intermixture as either a progressive or a dystopian scenario for humanity. The controversies between James Cowles Prichard and Josiah Nott, as well as between Arthur Gobineau and Anténor Firmin, are revised as regards the reproduction of different pathways of racialization. Exploring various methods for visualizing human variation complements the understanding of the longue durée of a Blumenbachian vision on human perfection as a significant diving force in the history of (de)legitimizing colonial power. The conclusion of the chapter problematizes the historicization of race in science and colonial power through the sociological biographies of scientists who have shaped our understanding of human diversity.

Keywords Race in science · Colonial power · Slavery · Abolitionism · Imperial (post-slavery) colonialism

3.1 Blumenbach's Legacy Within the Entangled History of Colonial Power and Race Science

In his 1940 tribute to the centenary of Johann Friedrich Blumenbach's death, Arthur Keith declared:

V. Shmidt (✉)
Centre for the History of Science, University of Graz, Graz, Austria
e-mail: victoria.shmidt@uni-graz.at

V. Shmidt and S. De Angelis (eds.), *Blumenbach's Racial Classification*, History, Philosophy and Theory of the Life Sciences 39,
https://doi.org/10.1007/978-3-032-15950-2_3

> These are some of the barriers which a century has raised between Blumenbach and us. But the chief has yet to be mentioned. Blumenbach accepted Genesis as authoritative; his successors seek elsewhere for guidance; everyone[1] of them has become an evolutionist (Keith 1940).

Keith's vision of Blumenbach as one of the key figures in emancipating evolutionism from preformationism resonates with Ruth Benedict's paraphrasing of the main points in *De Generis Humani Varietate Nativa* in her *Race: Science and Politics* (1940). Both Blumenbach and Benedict contrasted humans to other animals through "the power of reason and invention," (Benedict & Mead. 2019, 38) as well as humans' unique defenselessness (*man homo mitis et inermis*) (Blumenbach, 1775, 28), which transforms into the uniqueness of human survival and fitness. Moreover, while Blumenbach emphasized "slow growth, long infancy and late puberty, [which no other mammal has]" (Blumenbach ibid), Benedict stressed that in contrast to any animal who "would have grown himself a different body," "man's physical form remained stable." (Benedict & Mead ibid) Like Blumenbach, Benedict believed that the "[only] physical adjustment which may have survival value [for the human] is skin color." (ibid) Both applied an analogy between the lower classes and non-white races as demonstrative of human equality. Blumenbach's vision, which stipulated that " white men in a lower class rendered brown by a hard life; and it is equally certain that black men become whiter when they are less exposed to the effects of the weather," (Blumenbach, 1775 10) reverberated with the argument raised by Benedict in favor of asserting the equality between whites and non-whites:

> The peasants of Central and Southern Europe are as much a part of the White race as the aristocracy or the urban population, but their ways of procuring a livelihood, their narrow horizons in intellectual and social matters, are to the objective student less "civilized" than those of many Negro and Mongoloid "primitives." (Benedict & Mead 2019 55)

Blumenbach's approach to understanding the variation among humans united scholars with vastly different positions regarding racism, for example, Keith (who was convinced about the superiority of western Europeans) and Benedict (who was employed by the U.S. government and later by UNESCO to promote her vision of racial equality) (Brattain, 2007). And they were not exceptions among the anthropologists who referred to Blumenbach's legacy in the late 1930s.

The presence of a Blumenbachian approach is often understood to be on the verge of mysticism (Kühnast, 2020). In the shadow of WWII, the "good" abolitionist Blumenbach began to be opposed to "bad" German racial hygiene, especially by those who, like Keith, needed to emancipate themselves from recent close cooperation with German colleagues.[2] Further, the Blumenbachian five-race typology continued to infiltrate overtly racist approaches to the racial hierarchies that emerged in the

---

[1] The original orthography is kept.

[2] Karl Saller, the German anthropologist, a head of the Institute of anthropology in München (1948–1969), who was one of those charged with rehabilitating the reputation of German science after the fall of Nazism, viewed Blumenbach's legacy alongside the approaches of Gotthold Ephraim Lessing and Moses Mendelssohn: Saller, Karl Die Rassenlehre des Nationalsozialismus in Wissenschaft und Propaganda Progress Verlag Darmstadt 1961, p.21.

nineteenth century (see the chapter by Malin Sonja Wilckens in this volume). The institutionalization of physical anthropology that had started in the nineteenth century became an arena for disputes concerning evolutionary processes. Moreover, any approach alternative to Darwin, such as the holo-genesis proposed by Daniele Rosa, invoked the spirit of Blumenbach's idea of *Bildungstrieb* (the "formative drive"). Were the references to Blumenbach's legacy in such different directions mutually connected?

This question redefines the dilemma concerning the external vs. internal logic of scientific progress, through the lens of the topic as decisive for understanding how the weight of external and internal forces would be distributed (Turri, 2009). My revision of the ongoing adaptation of Blumenbach's approach to human variation for the last two centuries stems from understanding race in science as inherently connected to the vicissitudes of the legitimacy of colonial power.

Understanding colonial power as the dominance of the "center," over regions (and their populations) labeled as "peripheral" (Kohn & Reddy, 2024) allows us to recognize the core role of race-informed knowledge as a source for legitimizing and delegitimizing colonial power. The interrelation between colonial power and race in science becomes even more palpable when the history of colonialism is interpreted as a continuity of the crises of its legitimacy.

The history of colonial power can be seen as a continuity of several "ages," which have cemented the institutions of colonialism, including slavery, imperialism, genocide, autocracy, and racial capitalism. The transition in colonial power from one form to another is accompanied by a crisis in its legitimacy – when a predominant institution has lost its social recognition (Reus-Smit, 2007, 159). While the legitimacy of power is an inseparable part of the identities, interests, norms, rules, and principles that command social life (ibid 162), the loss of legitimacy is a complex process that should be seen as a driving force and an outcome of tectonic societal changes.

The obvious intractability of colonialism is rooted in the no less obvious capacity among agents of colonialism to reestablish its legitimacy. Targeted at reestablishing the acceptance of colonial power by its citizenry (Yoichi et al., 2013), the politics of legitimacy cry out for a justifiable vision of human difference and inequality. Race as "a set of logics attached to specific transnational practices" (Zimmerman, 2010) dovetails perfectly with this purpose.

The transformation of colonial practices has gone (and continues to go) hand-in-hand with advances in the study of human diversity.[3] Both those who have attacked one form of colonial power or another and its adherents have applied the outputs of scientific research regarding human origins and diversity to "prove" and advance their

[3] Mostly, the nexus of colonialism and race is explored regarding particular regions and local cases of colonialism; see, for instance: Gould, Stephen Jay. "The mismeasure of man." (1996); Howard-Wagner, Deirdre. "Colonialism and the science of race difference." In *TASA and SAANZ 2007 Joint Conference Refereed Conference Proceedings–Public Sociologies: Lessons and Trans-Tasman Comparisons, The Australian Sociological Association.* 2007; Ghoshal, Sayori. "Race in South Asia: Colonialism, nationalism and modern science." *History Compass* 19, no. 2 (2021): e12647.

positions. Scholars themselves have taken sides, offering either racialized or anti-racist arguments for supporting their political stance. Moreover, the entangled history of colonialism and racism is replete with examples in which an argument originally developed by a scholar who took an anti-racist position has been later used to support racist views or even to refute the anti-racist position. This is precisely what happened with Blumenbach's approach to human diversity – not only his classification of races but also the idea of *Bildungstrieb* has been recruited by those who seek to reestablish racial hierarchies for legitimizing imperial colonialism (Messling, 2016).

What was it about Blumenbach's approach that made this contradiction possible? Two extremes, either attributing to Blumenbach the role of an entrepreneur of racism (Gould 2014), or to defend the sustainability of Blumenbach's anti-racist position (Reill, 2018), seem to relegate the interrelation between colonialism and race in science to the margins of the critical historicization of science. A desirable critical narrative should embed Blumenbach, his scientific inquiry, and attempts at its dissemination in the context of colonial power and its vicissitudes during Blumenbach's time and afterwards. Such narration leads to not only to the task of historical justice for Blumenbach but also towards a more general question, namely, if racism is defeatable in terms of providing a sustainable biologically informed argument against the race-informed differentiation of humans.

Connecting the formation of colonial power with race in science proposes a more nuanced, or even a dialectical, view on the interrelation between scientific progress in understanding human variation and the politics constructed with the assistance of such arguments. One of the possible ways to avoid the risk of erecting a univocal determination between science and politics is to pay attention to interrelation between epistemic and ontological grounds for various scientific positions. The personal epistemology of Blumenbach is conducive to implementing this task. For the most part, modeling the epistemic and ontological cognition of Blumenbach remains rooted in the vicissitudes of the relationship between German Romanticism and its reflection in philosophy. Such a position embeds Blumenbach's legacy in the context of his complex relationship with his predecessors and, especially, Blumenbach's efforts to emancipate himself from creationism (Keel, 2018). It also leaves unanswered the question of how this aspect of Blumenbach's legacy shaped subsequent generations of scholars.

Visions of Blumenbach's approach to explaining human variation as either nurturing epistemically significant metaphors of human agency in evolution (Helbig & Nassar, 2016) or strictly positing a biological explanation opposed to a Kantian meta-cognitive use of *Bildungstrieb* (Richards, 2000) are complementary. These approaches present Blumenbach through the lens of "Modernity's Man," which stripped the human being down to instrumental rationality (Archer, 2003). But along with facing epistemological contradictions, Blumenbach apparently struggled with the conflict between the "death of God" and "titanic man" that framed the various approaches to examining humanity within the Enlightenment.

Phillip Sloan provides a nuanced view on the impact of the interdisciplinary exchange between Kant and Blumenbach, which resulted in an accurate historicization of their approach to the limits of understanding and its stimulating role in producing knowledge concerning human variation (Sloan, 2002). Through focusing on the consistent biologization of human perfection by Blumenbach, I aim at not only explaining the ongoing interest in Blumenbach but also discussing its role in the entangled history of colonial power and race in science. My research is inspired by two questions: How did Blumenbach problematize the history of humanity and human variation through the concept of human perfection? How has Blumenbach's conceptualization of human perfection affected subsequent generations of scholars and artists?

To revise the role of Blumenbach's approach in the first half of the twentieth century, I follow the transformation of his argument in three successive "ages" of colonial power: (1) the end of ***slavery colonialism***, in the last third of the eighteenth century; (2) the period of ***post-slavery*** or ***imperial colonialism***, mostly operating in the second part of the nineteenth century; and (2) ***post-imperial colonialism***, which started with the first crushes of empires (the Ottoman and Austro-Hungarian empires in particular) and continued until the end of the 1950s (finishing with the Algerian revolution). In this chapter, I focus on explaining the ongoing adaptation in the nineteenth century of Blumenbach's approach by anthropologists of both Western and non-Western origins as embedded in the process of legitimizing new forms of colonial power, namely, post-slavery colonialism in the age of intensive dissemination of imperial capitalism, which had replaced slavery colonialism.

In the following section, I redefine Blumenbach's approach to human variation as one of the conceptualizations that bridged the delegitimization of slavery colonialism and the legitimization of the post-slavery operation of colonial power, using a complex approach to human perfection. After discussing Blumenbach's temporal approach to human variation, I move to the consistent efforts by Blumenbach to make his argument regarding human variation more clear-cut through the visualization of human races.

Blumenbach's diagrammatic imaginary (Sommer, 2024 13) performed in three mutually resonating ways to illustrate the relation of races: (1) picturing the human skulls of four races as centered around the skull of the Caucasian, "perfect," race; (2) presenting comparable depictions of heterosexual pairs created by Daniel Chodowiecki; and (2) including the portraits of outstanding representatives of each race as an argument for race equality. Marianne Sommer highlights the nonhierarchical visual order developed by Blumenbach for explaining the anatomical affinity of different races (ibid 26). But which message regarding human variation was decoded in the illustrations by Chodowiecki for *Beyträge zur Naturgeschichte* (1790) and the choice of the illustration for *Abbildungen Naturhistorischer Gegenständen* (1796)?

The subsequent sections follow the adaptation of Blumenbach's approach to the aims of (de)legitimizing colonial power during post-slavery imperialism and the intensive institutionalization (and secularization) of physical anthropology, a main vehicle for race in science. These sections problematize the passage from post-slavery colonialism to post-imperial colonialism, through following the challenges facing evolutionary anthropology with "imposing continuity upon discontinuity and privileging ancestry over emergence." (Marks, 2012).

3.2 Human Variation and Its Perfection: The Essence of Blumenbach's Epistemology

3.2.1 Blumenbach's Abolitionism and the Temporality of Human Perfection

The incompatibility between slavery and humanism was fixed in European thought by Bartolomé de las Casas in the sixteenth century (Hanke, 2013).[4] However, only since the last third of the eighteenth century did the abolitionist movement challenge slavery, which started to be seen as an economic burden rather than a benefit, unacceptable among elites.[5] The Enlightenment and its actors played a crucial role in eroding the substantial support for slavery from the side of the citizenry.[6] The testimonies of scholars who had visited colonies, testifying a non-humanistic approach to African slavers and challenging stereotypes regarding Africans as inferior, were widely circulated in Europe (Rönnbäck, 2012.). For instance, the publications by the Swedish naturalist Andrew Sparrman, recruited by the British government gather information on slave populations, were translated into several languages, including German, and Blumenbach was certainly aware of its existence (Böttiger, 1795).

Abolitionism largely stemmed from "religious sources ... that made slave emancipation a symbolic test of the efficacy of Christian faith." (Davis, 1987) However, this initial motivation could not ensure the political success of the abolitionist movement. In search of a more convincing argument, the leaders of the abolitionist movement turned to economic justifications, which resulted in new forms of colonial power, namely, imperial capitalism, usurping abolitionism (Wright, 1973). This secularization of abolitionism through introducing expertise regarding racial equality aligned with the call for secularizing anthropology. Even in l'*Encyclopédie*, by de Diderot and d'Alembert, anthropology had been purely understood as the domain of theology:

---

[4] There have been recent attempts to situating the roots of the delegitimization of slavery in antique thinking; see, for instance, Ramelli, Ilaria. *Social justice and the legitimacy of slavery: The role of philosophical asceticism from ancient Judaism to late antiquity*. Oxford University Press, 2016. Further, the impact of the antique intellectual movements on modernity remains debatable; see, for instance, Alston, Richard. "Rereading ancient slavery." *Reading Ancient Slavery. London* (2011): 1–33. Finally, recognizing the impact of mainstream antique thought on practicing colonial power is much more widely disseminated; see, for instance, Hall, Edith, Richard Alston, and Justine McConnell, eds. *Ancient slavery and abolition: from Hobbes to Hollywood*. Oxford University Press, USA, 2011.

[5] The long-term debates among historians include different explanations for the unprofitability of slavery, for example: Woodman, Harold D. "The profitability of slavery: A historical perennial." *The Journal of Southern History* 29, no. 3 (1963): 303–325; Vedder, Richard K., and David C. Stockdale. "The profitability of slavery revisited: a different approach." *Agricultural History* 49, no. 2 (1975): 392–404.

[6] One of the most solid accounts of the interrelation between the abolitionist movement and the Enlightenment can be found in Drescher, Seymour. *Abolition: a history of slavery and antislavery*. Cambridge University Press, 2009.

> [The] Manner of expression, by which the crafty writers (sic!) attribute to God parts, actions or affections which are only suitable for men, and this to accommodate and proportion themselves to the weakness of our intelligence: thus it is said in Genesis, that God called Adam...". (Encyclopédie 1751 497)[7]

Historians agree on the fact that only in the first half of the nineteenth century the anthropological method had finally "escaped the abstraction of philosophical reflection and became a research method that fed the desire to convey positive knowledge." (Barbillon, 2004, 101).

Being among those who shared the ideology of and motivations underlying abolitionism, and who shaped the bases of anthropology, Blumenbach clearly faced these challenges of multiple secularization. In his letter to Johannes Loretz, one of the leaders of German pietism, Blumenbach declared his long-term devotion to the mission of proving that all people belonged to "one and the same race," stressing that his purely scientific method (*bloß naturhistorische Untersuchung*) "perfectly fits what the Holy Scriptures teach."(Blumenbach, 1791 76) The concept of the formative drive (*nisus formativus* or *Bildungstrieb*), central to Blumenbach's approach to human variation (more in a chapter by Simone De Angelis, in this volume), should be understood through the lens of this multiple secularization.

In *De Generis Humani Varietate Nativa*, Blumenbach (1775, 181) consistently deepens our understanding of humans as purchasing the "unlimited liberty of diet and locality" and portrays the human organism as perfectly equipped to "be able to live in every region under the sun." In today's parlance, we can define this conceptualization of perfection as an equalizing force among humans, who possess the unique ability to gain agency due to their outstanding physiological characteristics that adapt to new environments. Blumenbach's modification of the animated organism as agency, originally introduced by Buffon, emphasizes the formative drive as "the mother []... of varieties," the "generous origin of degeneration," or the "deflection from its usual path." (Richards, 2000, 20–21) Blumenbach primarily explained this deviation through the migration of humans to localities with different climates: "[T]here is almost no diversity of habit which cannot be produced by varieties of climate." (Blumenbach, 1775, 7) For Blumenbach, the diversity among humans represents the output of human sensitivity and responsiveness to such changes: "The flexible softness of ... mucous membrane in man is owing his power of accustoming himself more than every other mammal to every climate." (ibid 181) In this turn, human diversity is not only a direct consequence of perfection, but also its authentic performance.

Along with explaining the unique physiological features that manifest the perfect ability to adjust different environments, Blumenbach takes further steps in connecting perfection to adaptability. According to Blumenbach, humans are perfect, in contrast to animals, because they have "more nerves and more reaction." (ibid 179) Blumenbach emphasizes that prehistoric or, in his terminology, "old man" did not have

---

[7] The translation is the authors. See more regarding the discussion between French philosophers aimed at defining the limits of knowledge about the human body, anatomy and physiology: Sempère, Emmanuelle. "Le simulacre dans l'Encyclopédie: discours critique et anthropologie chez Diderot et Jaucourt." *Lumières* 1 (2018): 11–26.

the same sensitivity as "modern man," which makes prehistoric "man" similar to animals and less perfect. Not only do humans but also climate changes: "[W]e must allow something to climate, so far as that itself is altered by the lapse of time." (ibid 104) This point only increased Blumenbach's intention to describe the difference among races as a relational characteristic that depends on "how large and expanded the sample," which allows us to recognize "how similar look the people of North when they are compared with the people of South in contrast of comparing different subgroups of people of the North."(ibid 99).

This multilevel and mutually interrelated similarity among different groups led to claims of equality among various races, due to the vitality of racial intermixtures. Remarkably, Blumenbach built his argument as an invalid hypothetical syllogism by saying that if the hybrids would be not natural "the Supreme Being foresaw these disgusting kind of unions and took care to render them futile." (ibid 110) Blumenbach's reflections on godly and ungodly mixes were operationalized by Paul Broca (1864), who introduced a typology of human hybrids based on the degree of their vitality and the inevitability of their extinction. This historical continuity allows us to recognize the proto-eugenic thinking in the equalization of racial intermixture and the creation of humanity introduced by Blumenbach. The proto-eugenic sentiment becomes even stronger in Blumenbach 's analogy between the wild state of children brought up not as humans but as animals and disseminated diseases or defects like albinism—these causes do not shape self-reproducing groups but demonstrate multiple trajectories of deviations from "normal," perfect, ways of human development (Blumenbach, 1775, 130–133).

The duplicity of human perfection, based on the idea of perfect humanity and the perfection of particular humans, was key for understanding Blumenbach's ability to avoid binary oppositions such as preformationism vs. epigeneticism, fixism vs. transformatism (Palti, 2005; Dagg & Derry, 2024), empiricism vs. rationalism, and mechanism vs. vitalism, once defined as "multiple methodological parallelism." (McLaughlin, 1982) Bringing forward such a complex understanding of human perfection, Blumenbach encountered a question that intrigued him most of all, namely: How does a human being change over time, when humans exercise the freedom to explore the world? Do humans become more perfect or, on the contrary, do they lose their potential as agents?

This question marked a radical shift for Blumenbach, from his predecessors Haller, Linné, and Buffon, and led him to move beyond the limits of the multiple dichotomies that structured their scientific search. In Blumenbach's epistemology, life began to be described not only in terms of processes and phenomena in space and time, which substantiated the birth of biology (Stafleu, 1967, 432), but also through mapping the possible temporal dimensions of his idea of degeneration. Blumenbach's temporal vision on human perfection developed through his multiple reflections on aging, diseases, and deviations, which allowed him to provide a highly nuanced view on the difference between imperfection and degradation (see more in the chapter by Kseniya Brailovskaya in this volume).

This epistemic platform for developmentalism was further disseminated when Blumenbach's approach entered into polemics with another famous, representative figure of the Göttingen school of history, Christoph Meiners, who defended slavery primarily through claiming the inferiority of "Negros."[8] Meiners approached the difference between races and human groups and questioned human perfection through consistent application of a view on humans as structures of their collectives (Douherty, 1990). This relegated Meiners to attribute the main role to the location of humans as a main factor of differentiating humans and their progress (Wunderlich, 2021).

If Blumenbach operationalized human perfection through temporalizing extremal diversity in performing human agency, Meiners focused on the idea of human belonging to a particular group, pinned to particular location, as a predisposition of either perfection or imperfection: "[N]either a plant, nor a mineral, nor even a human being, loses the perfection (*Vollkommenheit*) ascribed to it by God when moved to other conditions." (Meiners, 1790) Further, bringing together animated (including humans) and unanimated nature should not read as a physicist or pro-antique position: Meiners was targeted by the intention to present locality as a universe. He brought forward numerous ethnological surveys in order to prove his stance and support his idea of beauty as a canon that stood in opposition to perfection understood as authenticity (Dougherty, 1990). Opinions regarding Meiners' theorization varied widely, from acceptance of his historicization as "being fantastic" because of the best possible access to sources for developing a racial hierarchy to labeling Meiners as a pseudoscientist (Michael, 2021).

The debate between the Jakobinian members of the Göttingen school, including Johann Georg Adam Forster, Georg Christoph Lichtenberg, and Blumenbach, and the adherents of slavery, Meiners and Samuel Thomas Sömmering, unpacked the issue of perfection, its possibilities and even its measurement. A Jakobinian and a friend of Blumenbach, Forster, directly attacked Meiners through employing a Goethean model of science based on radical empiricism. He built his argument on the disclosure of Meiners's idea of beauty and ugliness as "one when sufficiently attentive can also find traits enabling favorable and even pleasant portrayals of very many, even of the wildest, peoples, written up by the most credible authors." (Gray, 2024).

The analogy between humans and plants, adapted by Forster from Goethe's view of the *Urpflanze* ("archetypical plant"), aimed at proving the ability of humans to be re-rooted, a proof that reverberated with Blumenbach's vision on the incredible adaptability of humans but obviously challenged his devotion to human agency.

---

[8] Meiners published an enormous number of pamphlets aimed at promoting his view on slavery; the most disseminated were two articles entitled "Historische Nachrichten über die wahre Beschaffenheit des Sclavenhandels und der Knechtschaft der Neger in West Indies" [Historical Report on the True Quality of the Slave Trade and the Servitude of the Blacks in the West Indies], and "Von den Varietäten und Abarten der Neger" [Of the Varieties and Deviate Forms of Blacks], which were published in the *Göttingisches historisches Magazin* in 1790, *Christoph Meiners: Ueber die Natur der Afrikanischen Neger, und die davon abhangende Befreyung, oder Einschränkung der Schwarzen.*

Additionally, Meiners was not entirely alien to the idea of human agency, which he promoted in his consistent critique against the aristocracy as opposed to the people, in order to stress the unique suitability of the bourgeoisie to the "modern" order of social life (Golf-French, 2019, 946).

The irony regarding the nobility was not alien to Blumenbach either. According to Mario Marino, Blumenbach explained the generous gesture by the British King to present the skull of Richard Bruce King of Scotland to Blumenbach's collection through the fact that the king did not need this object of national pride any longer (Marino, 2021 75). Also in contrast to Blumenbach, Meiners refused the possibility of establishing a social order with the predominance of enlightened, middle classes among non-whites.

Beyond the differences and similarities with Blumenbach, it is worth mentioning that Meiners's approach warranted consistent positive acceptance, especially among those affiliated with authorities (Benes, 2023). In this turn, it is reasonable to interpret the methods of visualizing human variation chosen by Blumenbach as determined by the intention to depict a desirable post-slavery social order, opposed to pro-slavery approaches like Meiners's. *Beyträge zur Naturgeschichte*, which Chodowiecki was invited to illustrate, were published in 1790, the year when Meiners published his article defending slavery, along with *Abbildungen Naturhistorischer Gegenständen* six years later.

3.2.2 Visualizing a Variety of Races: Towards the Patterns of the Post-Slavery Social Order

> The trees are planted in rows, not so densely. The people are not wild and shaggy, but elegant in their own way. The men only have a short beard and a large mussel shell hanging from a coral necklace. [They wear] shorter aprons than the women (Blumenbach, 1782/2006, 187).

These instructions were received by Daniel Chodowiecki from Blumenbach, for improving the image of Australians and Polynesians after sending him the first sketches of the images for *Beiträge zur Naturgeschichte.* In formulating his expectations, Blumenbach heavily relied on multiple sources about the colonies collected by Europeans (Hünniger, 2018), which corresponded with the canon of the close gaze of natural scientists established in the eighteenth century (Bleichmar, 2007). Further, in his intention to represent human dignity, Blumenbach was aligned with a canon recently established by pro-abolitionist artists that showed noble "savages" with their own sense of beauty (Mielke, 1988), who started to be depicted with babies, fishing or hunting instruments, or local animals.

Moreover, it is reasonable to move beyond an examination of these illustrations as pure adaptations and combinations of already rooted stereotypes regarding "exotic" races and localities. As is well known, underlying the invitation by Chodowiecki to illustrate the book was Blumenbach's uncertainty about the quality and the sufficient description of the objects through using images from the university collection; he thus decided to make totally new pictures (Collet, 2014, 254). But what ensured mutual understanding between the scientist and the artist?

Along with Chodowiecki's closeness to the Göttingen school and its Jacobin camp in particular,[9] the decisive factors included sharing the experience of in-betweenness regarding social belonging: both men originally came from lower-middle-class families and had achieved their success through the never-ending fight to be accepted. By the time he met Blumenbach, Chodowiecki had garnered the reputation of not only a responsible and successful etcher who illustrated numerous popular books, someone able to connect different localities and to nuance their visualization (Schultz &Wróblewska, 2020). This sensitivity of the artist was crucial to the clear educational purposes of *Beiträge zur Naturgeschichte.*

If the main intention behind including images was enlightening and promoting the idea of equality, it is reasonable to assume that Blumenbach and Chodowiecki were connected by a secular, humanistic concept of *Bildung*, which had begun to shape public space in Germany in the eighteenth century (Kontje, 2019). This movement launched the transformation of the position of those who should be educated toward more proactive interaction with the environment. Aligned with this, education and its outcomes, namely, good or bad customs, began to be seen as "second Nature." (Shapin, 2019) Chodowiecki, who was extremely devoted to natural approaches and encapsulated his view in a series of cartoons aimed at contrasting the lives of naturally and artificially educated men and women, could easily share the Blumenbachian view on the differences among the races as a product of different but equally good "second natures." (Lichtenberg, 1901).

Naturalness assisted with the choice of how to demonstrate the similarities in races—picturing each of the five races as a family unit. This intention was close to the artist's devotion to portraying attractivity of family intimacy. Chodowiecki created numerous images of happy families of different class backgrounds, but always through the lens of idealizing the middle class in opposition to corrupted aristocratic elites (Ehler, 2003). For the vignettes in *Beiträge zur Naturgeschichte*, Chodowiecki presented each of the races by a couple in their prime, at the peak of their reproductive abilities. This choice corresponded with *calendar time,* an approach to portray daily life that brings together "the cultural representations of the past [defined by] the vague category of 'myth,' and all the regulated and recurrent symbolic practices [placed] under the no less vague name of 'rite.'" (Calame, 2009) This composite temporality assisted in the task of balancing the similarities and differences among races. Images of five races combined deliberate signifiers of civilization for the races, seen by Blumenbach as the most distanced from Caucasian, and the multiple signifiers for grading the differences between the Caucasian and other races (more similarities with Asians and less with the rest of races).

[9] By the time of the cooperation with Blumenbach, the artist had already collaborated with Lichtenberg, who had recruited the artist to illustrate Göttinger Taschen-Kalendat; more in Claire Baldwin Authority and interpretation in G.S. Lichtenberg in Poag, James F. and Claire Baldwin. The Construction of Textual Authority in German Literature of the Medieval and Early Modern Periods. The University of North Carolina Press, 2001.

If Caucasians were depicted in the close environment of a house, Asians were in a semi-closed gazebo, Africans, Americans, and Australians/Polynesians were depicted in open space; even their housing was part of the background. The imaginaries of the background landscapes on the portraits of all the races except the Caucasian were full of fecundity symbols. Caucasians and Asians were depicted at the highest moment of mutual sexual seduction, but Africans, Americans, and Australians/Polynesians were pictured as families with children. Africans were shown with the youngest children, being breastfed, Americans are pictured with toddlers, and Australians/Polynesians are depicted with 6–8-year-old children. This interrelated difference reflected pre-Darwinian evolutionary thought, in which was typical to present "entire peoples and races … as part of the childhood of the human race, in need of guidance and protection certainly, but with the potential (however distant in prospect) for achieving the adult state." (Steedman, 1995, 82–83) Practically, by emphasizing the cyclical components of biological time for different races, Chodowiecki cemented the Blumenbachian approach to racial hierarchy.

His cooperation with Blumenbach only reinforced Chodowiecki's reputation as one of the pioneers of establishing a new standard of human perfection "associated with the normative view of culture turned out, in practice… [which had] strong connections with the particular views of ruling groups and classes" (Bennett et al., 2013, 67) Furthermore, the artist's engagement in promoting the independence of the United States guaranteed the acceptance of Chodowiecki as a creator of the canon of Europeanness and its intensive adaptation by American painters (Maierhofer, 2023). Even in the 1920s, he continued to be seen as the artist of his time, and not as a timeless genius (Greenberg, 1921). The long echo can be accepted as one of the driving forces behind the adaptation of Chodowiecki and Blumenbach's approach to visualizing human variation by subsequent generations of anthropologists. For Blumenbach, the task of finding additional suitable methods for visualizing human diversity remained important.

The period between the publication of *Beiträge* and *Abbildungen* saw the most intensive collection of skulls for Blumenbach's collection (Böker, 2018, 9). New additions were accompanied by intensive correspondence with fellow doctors, who sent Blumenbach skulls from different parts of the world. This correspondence motivated Blumenbach not only to revise his criteria and approaches to describing races and the differences among them but also to redefine his approach to informing the public about races.

In his *Abbildungen Naturhistorischer Gegenstände*, Blumenbach had chosen the portraits of five famous non-Europeans men, who had achieved acceptance among Europeans and high positions in societal hierarchies, for illustrating five main races: the Mongolian painter Feodor Iwanowitsch Kalmyk, the Mohawk leader and diplomat Thayendanegea, the Ottoman ambassador to Britain Jusuf Efendi, the celebrity O-mai, and the writer and clergy Jakob Capitein, whose personal achievements are described in lieu of "racially" typical traits (Eigen, 2005, 285). By selecting successful, powerful men for representing the variety of human races, Blumenbach created an additional signifier of human capital, achievable for different races: the opportunity to be pictured by famous artist or even by themselves (Kalmyk was presented through a self-portrait). Each of the portraits was accompanied by a description aimed at demonstrating the outstanding progress towards the highest standards of being civilized:

> I consider it less a proof of [O-mai's] docility that he made admirable progress in chess in a short time, than that he very soon adopted the refined London worldly tone so well that the famous Dr. Johnson, when he once dined with him in company and O-Mai sat next to Lord Mulgrave opposite the doctor at the window, so that the light blinded him, he, according to his own confession, could not distinguish him from the Lord at first, due to the elegance of O-Mai's manners (Blumenbach, 1796, 36-37).

Bringing forward the social success and acceptance of men of different races assisted with overcoming, albeit partially, the contradiction between individualized humanism and the obligation to describe races as "a group of people who exhibit a defined set of characteristics distilled from a finite sample group." (Eigen ibid 280).

Such an approach to depicting human variation calls into question the widely disseminated, generalized view on illustrations that accompanied racial hierarchies as aimed at embodying population groups through individual portraits (M'charek & Van Oorschot, 2019) Blumenbach's position created the preconditions for the subsequent resolution of the dichotomy between the collective (spatial) and individual (temporal) approaches to racial diversity in the next age of colonial power, when the role of class in racial hierarchies became decisive after delegitimizing slavery.

3.3 Blumenbach's Legacy During the Age of Post-Slavery Colonialism

Post-slavery colonialism gained momentum during a period of multiple transitions in scientific understanding of humanity: from Enlightenment to positivism, dissolution of the monogenist-polygenist debate in the polemics around Darwinian evolutionism, and through a major reconsideration of the relationship between the social and natural sciences to better understand humans. The multiple revolutions were accompanied by interstate strife over colonial territories, which made the legitimization of colonial power increasingly dependent on the loyalty of citizens to the integrity of their empires. Within the context of these social and political shifts, Blumenbach's vision on human diversity and perfection gained new possibilities for adaptation and transformation. One of the decisive factors in adapting Blumenbach's vision was the relationship between slavery, race in science, and abolitionism, which became more convoluted as the abolitionist movement progressed.

3.3.1 *Prichard's Effectuation of Blumenbach's Vision of Human Variation*

In the "the longer shadow of the ending of the slave trade," (Drescher, 1990, 445) which continued throughout the nineteenth century, Blumenbach's argument remained desirable for abolitionists, both those devoted to the mission of demonstrating the ideals of Christianity and those who derived their arguments from natural science and positivist philosophy.[10] Blumenbach's monogenism acted as an arbiter between creationism and teleomechanism, producing a wide range of pathways towards his approach to human variation and the driving forces behind it. The close and amicable relationship between Blumenbach and the German Pietists strengthened the circulation of his approach, not least because of the relationship between German Pietism and Quakerism in New England (Jantz, 1942). *This flow was connected to one of the first appropriations of Blumenbach's vision of human perfection, which occurred within the framework of the collaboration between William Blake, who was in direct communication with Chodowiecki, and Johann Caspar Lavater. The* long echo of Blumenbach's perspective among religiously oriented abolitionists and their art should be seen as one of the factors underlying the ongoing tradition of presenting Blumenbach's legacy as a part of the history of anti-racism (Rainger, 1980).

Along this dimension, Blumenbach' authority can be easily attributed to imperial humanitarianism, a broad category of philanthropic endeavor with anti-slavery sentiment at its core (Reid, 2024) and pinned to a range of diverse geopolitical anxieties among European empires. Was abolitionist thinking, including the adaptation of Blumenbach's approach, exclusively a part of "self-congratulatory rationalizations"? (Drescher, 1990, 444). How heuristic was Blumenbach's conceptualization of perfection for producing knowledge about human variation? Answering these questions must involve the sociological biography of James Cowles Prichard, one of the founders of the Ethnological Society, a quaker physician, and the author of *Researches into the Physical History of Man Dedicated to Blumenbach*, published in three editions between 1813 and 1847.[11] Along with deepening his argument in favor of monogenesis and developing a vision of human history, Prichard remained faithful to Blumenbach's holistic approach to human perfection:

> Nature seems... to be every-where satisfied with resemblance. Her purposes are attained by displaying that sort of general analogy which is still compatible with individual variety. It

---

[10] One of the most solid mappings of the diversity of epistemic platforms disseminated among abolitionists can be found in Haller, John S. "The Species Problem: Nineteenth-Century Concepts of Racial Inferiority in the Origin of Man Controversy." *American Anthropologist* 72, no. 6 (1970): 1319–29.

[11] The last, third edition, was issued in the period between 1836 and 1947 and consisted of five volumes. The attempt to provide a solid historical analysis of Prichard's legacy was implemented by Hannah Franziska Augstein in *James Cowles Prichard's anthropology: remaking the science of man in early nineteenth-century Britain.* Vol. 52. Rodopi, 1999. Furthermore, there are some factual mistakes in the monograph, which presents a question as to the quality of interpretation.

> is most probable, that no two individuals were ever produced in any species with complete and perfect sameness of form and structure (Prichard 1836, 242).

In addition to this vision of perfection as a general idea about human variation, Prichard shared research questions that guided Blumenbach, particularly in examining the trajectories and drivers of human change and how these changes were reflected in the diversity of races. Two methods for exploring human variation were offered by Prichard, "ethnographical" (investigation of phenomena occurring among the particular races to be compared) and "analogical" (investigation into the general laws of their economy). These methods equipped him to "determine what in the structure of species is permanent, and what is 'able to transmutation.'" (ibid 110–111) The distinction between ethnographical and analogical methods transformed the opposition between temporal and spatial approaches to human variation into the complex task of temporalizing the space of human inhabitation. And for humans such location includes the entire universe, according to Prichard.

Prichard was open about the numerous blank spaces within the knowledge of both external space and the organization of living bodies, which developed into an argument in favor of a functional approach to studying human diversity (ibid 112). He highlighted what he marked as "vital functions": the periods and duration of life, the economy of the sexes, the phenomena of parturition and reproduction (ibid 114). Prichard moved away from the classification of races introduced by Blumenbach in favor of a more complex and ramified classification. This positioning turned him away from the teleomechanical approach of Blumenbach. In short, Prichard not only moved away from the concept of *Bildungstrieb* but also abandoned thinking about humans in terms of agency; according to him, humans were "animals of the same species … subjected to nearly uniform laws with respect to all the principal functions of the animal economy." (ibid 114) Moreover, the sensitivity and responsiveness of humans central to the Blumenbachian conceptualization of human agency was transformed into human educability, equal for all races, according to Prichard.

Along with multiple comments regarding the hypersensitivity of the races seen as inferior as savage nations, or those who live by hunting (ibid 264), Prichard focused on the "reception and progress of Christianity" (ibid 187) and "conversion to civilized habits" (ibid 196) as the main markers for demonstrating that the minds of such races bear the same moral and intellectual constitution as that of other human beings. By deepening this argument, Prichard significantly complicated Blumenbach's comparative method, instead pioneering and practicing a method which contemporary methodologists might define as the "method of difference":

> There are no races of men so widely differing from each other in structure of body, and in physical properties as the three above mentioned [the Esquimaux, the Hottentots, and the Negroes], and the fourth class of nations [Europeans] with whom they have been throughout compared: if therefore the principal conclusion is allowed with respect to these instances, it will scarcely be thought disputable in others (ibid 215).

Remarkably, the second edition of his *Researches into the Physical History of* Man published in 1826 opened with a portrait of Abba Gorgoryos, an Ethiopian priest of noble origin, -which can be interpreted as adaptation of Blumenbach's approach to present the variation among human races through portraits of outstanding representatives of each race.

Prichard further described the multiple performances of the "prevalence of superstitions and practices more or less resembling the Fetissism[12] of Africa" as existing among "nations enjoying a much higher degree of mental culture," adding that "such are a belief in destiny or fatality, astrology, necromancy, charms, spells, omens, lucky and unlucky days, fortune, and the good and evil genius of individuals." (ibid 197–198) Later, this dimension would be developed by American anthropologists Margaret Mead, educated in a Quaker family, and her close colleague, Ruth Benedict, who posited Prichard as one of the most consistent antiracists, along with Jean Finot and E. A. Hooton (Benedict & Mead 1940, 36).

Prichard continued reproducing Blumenbach's understanding of human history as the degeneration of races, but he moved away from attributing the historical role of the cradle of European civilization to the Caucasus (Prichard ibid). In his last edition of *Researches into the Physical History of Man*, he developed the hypothesis of the Fertile Crescent, which pointed to the Middle East as the cradle of Western civilization, achieving its heyday at the beginning of the twentieth century in the texts by James Henry Breasted (Ambridge, 2012). Prichard's detailed description of Iranian (Indo-Atlantic) and Turanian nations as the two main groups necessary for understanding the inception of human civilization was remarkable due to a consistent focus on intermixing between these two groups. One of the predominant motifs of the latest edition of the book included a very outspoken position in favor of understanding racial intermixture as a pathway toward human progress.

The third edition included an entire subchapter entitled "On mixed human races," in which the main leitmotif was the superiority of intermixed races, in terms of both physical and mental abilities. Even more, Prichard stressed that their fitness permitted mixed races to obtain higher social positions and to replace those races that were the predecessors of such mixing: "[T]he mixed offsprings originating from different races within the limits of the same specie generally exceed in vigour and in the tendency to multiplication the parent races from which they are produced." (Prichard 1836, 150) With this radical acceptance of racial mixing, Prichard pointed to an equally radical resolution of the dichotomy between spatial and temporal approaches to the study of human variation. He cited several examples of racial mixing around the world that had acquired a profitable animal economy, unlike their parent races. For Prichard, this diversity gave racial mixing a position as one of the vital functions crucial for explaining the unity of humanity.

The fact that Prichard discussed racial intermixture as the only way toward human progress, in close connection with the fear of the extinction of humanity, reflected the context within which Prichard elevated the role of racial intermixture. In his own way, Prichard was responding to the heated debates in English society between 1835 and 1836, when a U.S. clergy had introduced the issue of "amalgamation" (the definition used in the United States for describing interracial mixing), to the British public. Presenting racial intermixture as an inevitable path toward human extinction

[12] The author's orthography is kept.

was fixed in popular dystopian novel, *A Sojourn in the City of Amalgamation, in the Year of Our Lord, 19–*, written by Jerome B. Holgate under the pseudonym of Oliver Bolokitten; it presented the death of U.S. society due to mixing of white and Black people (Murison, 2011). It is not clear if Prichard was aware of the novel but undoubtedly, he could not overlook the debates in British newspapers, for instance published by the *Patriot*[13] in the summer of 1836, which presented the British public with an option to maintain its Chrisitan identity, but, at the same time, to practice polygenism and to stand against a racial "mosaic." (American Slavery, 1836) Prichard's "On mixed human races" can be easily read as a direct reply to such ideas:

> God needs very race that now peoples it; and must see the hand of God in arresting so speedily … this pernicious heresy [mixing races]. This prejudice of colour…was in all respects mutual. …[I]t was the common instinct of the human race, and existed as really… on the side of the coloured population as on that of the whites; …those who attempt to promote amalgamation are fighting equally against the purposes of Providence, the convictions of reason and the best impulses of nature (Prichard, 1836, 267)

The polemics initiated by Prichard, from both scientific and moral positions, gave a long life to the association between an anti-racist stance, including the acceptance of racial intermixture and a racialized view, with the struggle for racial purity. The attack against Prichard's stance from the side of Joseph Arthur de Gobineau cemented this division.

3.3.2 *The Mutation of Blumenbach's Argument in Post-Slavery Racialization*

In his *Essai sur l'inégalité des races humaines* (*Essay concerning the inequality of human races*), Gobineau denied the idea of the educability of inferior races, along with racial intermixture, as prerequisites for human progress. He brought forward the case of the Jesuits and their attempts to "civilize" the natives of Paraguay, who were "[treated] like children… [under] a despotic government of Europeans, and who felt once more the call of the savage life with the exception of thirty-seven straggling little villages… in which the population is, no doubt, *partly hybrid* (emphasis mine)." (Gobineau 1853, 53 53).

Gobineau discussed the historical role of "half-breeds," more aligned with Alexis de Tocqueville, whose *Democracy in America* was referenced as a counterargument against Prichard's position:

> The half-breed shares the enlightenment of his father without entirely giving up the savage customs of his mother's race. He is thus a natural link between civilization and barbarism. Wherever half-breeds exist and multiply we see the savages gradually changing their customs and social conditions (ibid 72).

[13] *Patriot* was established in the early 1830s by the group of clergy and religious experts interested in consolidating people around an updated religious identity. The newspaper quickly enjoyed a wide audience; more in Cooper, J. Nicoll. "Dissenters & National Journalism: 'The Patriot' in the 1830s." *Victorian Periodicals Review* 14, no. 2 (1981): 58–66.

Furthermore, already existing "half-breeds" were included by Gobineau among the distinct families into which humanity could be divided (ibid 108). This view on "half-breeds," as not embodying a promising future but not necessarily negatively influenced by the past, sheds light on the complex interrelation between Gobineau's dystopian view on humanity as inevitably in a state of degeneration, his flirting with the disappointment in aristocratic elites, and his consistent position on racist anticolonialism (Blue, 1999 118).

Gobineau's division of the history of humanity into "the earliest epoch… when the influence of natural causes was far more active… immediately after the Creation, when the earth was still shaken by its recent catastrophes and without any defense against the fearful effects of their last death-throes [and] still continuing now… under the normal and healthy conditions," (Gobineau 1853, 137) approximates Meiners's exemplary historicization with its inevitable objectification of humans.[14] This anti-agential thinking was heavily reinforced by a consistent application of comparisons with animals for justifying the existence of racial hierarchies.

While reaching its peak, the struggle between pro- and anti-slavery forces resulted in paradoxical outcomes regarding the interrelation between scientific and religious arguments about human origins. While abolitionists grew increasingly frustrated with the need to secularize their arguments, pro-slavery advocates began to seek ways to revise biblical teachings regarding the origin of humanity. Nott, like other proponents of polygenism, nurtured the theological origins of his theory and turned to the teachings of Isaac La Peyrère, a French priest of noble origin, whose rejection of biblical arguments for the unified origin of humans was given a second life in the first third of the nineteenth century.[15] The pre-Adamites hypothesis began to be directly opposed to an understanding of the Bible as propagating monogenism, according to Blumenbach and Prichard (Philalethes, 1964). When criticizing polygenism as religious heresy started losing its legitimacy, polygenism obtained a more solid position and began to be accepted as not being contrary to Christianity. Such a position partially explained the obvious success racially minded polygenists obtained, not in protecting slavery but in establishing the long pathway of racialization through attributing inferiority as a consequence of racial intermixture.

Prichard's approach, including his reliance on Blumenbach's concept of perfection, faced massive criticism from the side of those who defended slavery from the position of polygenism, including Josiah Clark Nott, slave owner and surgeon in the United States, who, together with George R. Gliddon, claimed in *Types of Mankind* (1854) that racially mixed offspring are not sterile but demonstrate lower levels of fitness. The opus by Nott and Gliddon was framed (obviously with intention) as the consequence of the logical complements of Prichard's vision on human origin

[14] Gobineau recognized Meiners's division into "beautiful" and "ugly" races as well: Arthur de Gobineau The inequality of human races, p 107.

[15] Among overt racists, La Peyrère gained the reputation of a martyr because his work was forbidden to be disseminated. This heroization continued until the early twentieth century: Von Hofsten, Nils. "Zur älteren Geschichte des Diskontinuitätsproblems in der Biogeographie." *(No Title)* (1916). Livingstone, David N. "Cultural politics and the racial cartographics of human origins." *Transactions of the Institute of British Geographers* 35, no. 2 (2010): 204–221.

and variation. The book not only nurtured a polarizing view on truth and falsehood, so important for Nott, who was interested in public acceptance,[16] but also dealt with an issue the adherents of slavery faced – namely, the need to find a balance between their vision on humanity as made up of unequal races, and their belief in Christianity. This context inclines us to move beyond understanding the contest between Blumenbach and Prichard on the one side and Nott on the other side as simply pinned to one or another degree of inclination toward evolutionary undertones (Keel, 2013, 16).

At the core of Nott and Gliddon's attack against Prichard was that understanding that his method of analogy was unable to produce a univocal argument based on "simplicity and uniformity in the laws of nature" but rather generating casuistic argument "both for and against the unity of the human race." (Nott & Gliddon 1854, 17) Meiners' vision of races as pinned to localities was exaggerated:

> "[It was] the decision of the Allwise Creator regarding the best suited climate and situation to the organizations of different races …The black man was placed in Tropical Africa, because he was suited to this climate and no other. The white man was placed in Europe and Asia, for the same reason." (ibid 29)

Prichard's opinion regarding the absence of difference in vital functions was immediately questioned with the assistance of statistical data regarding the higher rates of death among free people of color in contrast with slaves, and even higher rates of mortality among "Mulattoes."(ibid 30).

Nott and Gliddon did not ignore the relationality of human races claimed by Blumenbach and Prichard, saying that "the best specimens of Negroes are very like the inferior of the Caucasian… but when you compare extremes the argument fails…" (ibid 24) The intersectionality of disability and race became explicit in the argument, which brought forward the size of brains among "idiots" and people of color as equally smaller than the brains of healthy white individuals (ibid 463). The illustrations emphasized the disproportionality of the non-white races and were chosen with the expectation of drawing an analogy not only between "lower" races and the humanoid apes, but also with images of races from the past - which, according to Nott and Gliddon, ostensibly proved the immutability of the inferior position of these races in the hierarchy. Blumenbach's argument about human sensitivity distinguishing humans from animals mutated into a hierarchy of races regarding the degree of humanity. Non-whites were noted as having thicker and less sensitive nerves and lower "intellect." (ibid 23, 461–462) The desirable racial purity of "Negros" could be achieved only through fixing their belonging to a lower class "and impossibility to mix with those socially higher whites," (ibid 16) a view that resonated with Gobineau's' argument against racial intermixture, although Gobineau himself was very critical of the adaptation of his theory in favor of defending the institution of slavery (Wright, 1999 833).

Converting Gobineau's pessimistic view on racial intermixture as embedded into an "unnatural" turn in human history regarding strategies of colonialism verging on

[16] Nott was well known as a propagandist of his racist views with help of public lectures which he published too: Nott, Josiah Clark. *Two lectures on the natural history of the Caucasian and Negro races*. Dade and Thompson, 1844.

genocide, made by Nazi ideologues, can be easily read as an attempt at making the lost dream of "proper" and "natural" order once again possible - through genocide (Biddiss, 1966). Clearly, this historical vicissitude only added to the points for interpreting the opponents of Gobineau (first of all, Blumenbach and Prichard) as the bulwark of anti-racism and adapting their agential vision.

3.4 Race Sentiment in Post-Slavery Emancipatory Movement: The Strange Case of Anténor Firmin

A critical reconsideration of Gobineau's approach was a solid part of the agenda among French anthropologists since the last third of the nineteenth century. In his short and alarmist comment, Paul Topinard built his argument against Gobineau's equalization of cultural and physical diversity as totally anti-scientific, recognizing its frighteningly stable influence on such movements as Pan-Germanism and Pan-Slavism, due to the role of diplomats. Remarkably, Topinard (1901, 11) forecasted (rightly) the ongoing reproduction of Gobineau's approach after empires died down. Being a student of Paul Broca, Topinard deepened Broca's approach to hybrids in terms of their vitality (Broca 1864), establishing a further criterion for racial hierarchy, namely, more and less adaptable mixes between superior and inferior races.

Experiencing on his own skin how discriminative such an approach could be (Bernasconi, 2008), Anténor Firmin, a self-educated anthropologist "located in the category of the Negro," (Gordon, 2018) developed his own account of racializing humanity. By acknowledging the equality of Blacks and whites as a prerequisite for a non-discriminatory approach toward racial intermixture (Firmin 2000, 204), Firmin connected Prichard's claim regarding the desirable future of humanity as racially intermixed with shared responsibility for such equality, also due to minimal predictability – in terms of biological features:

> In the initial instance of hybridization, each parent's hereditary physiological baggage intervenes in opposite directions but with equal force, each modifying the other and producing an offspring with the average characteristics of the two parents. In the second instance, the hereditary traits, already diffused and mixed, come into contact and combine in unpredictable ways.... (ibid 207- 208)

To achieve a "desirable" social order in a multiracial world, Firmin called for limiting the transfer of biological methods "to the political organization and management of human society [as destroying] all spontaneity and all progress." (ibid 312). Instead, he developed a complex interdisciplinary vision on anthropology, which later would be adapted by Franz Boas and other American anthropologists (Lewis, 2018) In this turn, Firmin took another step in the direction indicated by Frederick Douglass, another thinker of African descent who advocated environmental monogenism as a prerequisite for the further flourishing of the "American" nation (Douglas, 1999).

If regarding the formation of anthropological methodology, Firmin's approach manifests a revolutionary move toward cultural anthropology (Craipain, 2024), he was challenged by the task of deconstructing the expeditiously changing colonial order. In the midst of Firmin's argument, we can find the idea of the right to progress:

> In sum, there are savage nations and civilized nations. Naturally, the former are inferior and the latter superior. Race has nothing to do with it; civilization has everything to do with it. Despite the state of deep degradation in which those savage races are believed to be, they have not lost their right to partake in humanity's common patrimony (Firmin 285 286).

His conviction about mutually opposed historical trajectories of white and Black races relegate colonial power to the margins of driving forces behind "profound decadence" and attributed to the white "race" the responsibility for "complete bastardization" of Blacks (ibid 330). On the one hand, Firmin stressed the educability of Blacks and provided multiple examples of their achievements, often limited by the white prejudice already internationalized by Blacks themselves. On the other hand, he consistently emphasized the very limited number of whites (among whom Firmin mentioned Blumenbach and Prichard) who were willing to support the progressive intentions of blacks, in contrast to the vast presence of those who were racially prejudiced. Firmin, like Blumenbach, relied on intellectual elites; in fact, much of his argument regarding racial equality was operationalized through the many examples of the equality or even the superiority of people of color who had achieved education and status. This relational approach to connecting racialization and progress becomes even more distinct in Firmin's argument against the reduction of all explanations regarding the problems of "inferior" races to their lack of activity in solving the problem of adaptation:

> It could be objected here that many nations found themselves in the best of conditions for their development and yet remained in a patent state of inferiority, whereas others, less favored by circumstances, were able to march forward, moving from one conquest to the next until they reached the highest summits of civilization. (ibid 275)

This vision concerning the combination of the driving forces behind the perfection of some races and the imperfection of others certainly suggests that du Bois would apply to Firmin a critical argument (like that applied to Blumenbach), which recognizes a completely "natural," phenotypical conception of race (Harris, 2019).

One of the significant factors in the reproduction of the controversy between supporters and opponents of racial mixing includes the demand for such explanations during periods of nation building (Darlington, 1961, 19–20). While "Darwin was able to reconcile monogenism with the widely perceived fact of great human diversity and, though the mechanism of natural selection, to explain evolution as a long, slow process … it was not helpful at all in explaining national differences between, for example, the peoples of Europe." (Mandler, 2006, 73) The call for managing racial intermixture as desirable selectivity became even more evident in the late nineteenth century, when nations began to face the crush of empires (Mudford, 1968). The approaching collapse of empires and the growth of nationalistic tendencies throughout Europe created all the preconditions for the subsequent adaptation of Blumenbach's ideas and the ideas among his opponents, enacted by those who sought sustainability in institutionalizing anthropology at the transnational level (Baskar, 2016, 66–67).

Jacques Barzun quite sarcastically erased the boundaries between the camps of Blumenbach's supporters and opponents in favor of understanding their discussions as "ill-concealed animosities of nationalist underlaying the vigorous enmity of rival theories." (Barzun 1965, 125) Barzun, who critically observed and recorded the flourishing of post-imperial colonialism, when nation-states become the most vocal representatives of employing race in reinforcing their power, noted that "Turman and Barnard Davis, Van Duben, Pruner-Bey, Broca, Topinard, Morton, Nott, Gliddon and the shades of their predecessors or colleagues, Blumenbach, Cuvier, Prichard, Retzious, Rudolf Wagner, Huxley, Haeckel, and Virchow fought Homeric battles wanting only the laughter of the gods." (Ibid)

However, it was not only a new form of colonialism that made Barzun idiosyncratic with respect to previous generations of antiracists, but also his exclusively positivistic attitude towards science, in which, he opined, there was no place for the conceptualization of equality: "[I]t is neither provable not disprovable." (ibid xix) For Barzun, post-imperial colonialism gave race the space to operate as a principle of political and emotional unity. For liberally thinking Barzun, unity based on race ignores the variation among groups and social interests in favor of inevitable genocide: "Perhaps, the new nations require this artificial welding, but they will regret its costly efficacy when it leads, as it must, to mutual massacre." (ibid xx).

3.5 Conclusion

Since the abolitionist movement achieved its goals, colonial power began to exist in the regime of an ongoing fight for its legitimacy, and race science obtained a central position in this fight. Blumenbach and his network were among the first entrepreneurs in the crisis of colonialism's legitimacy. Blumenbach was one of those who laid the foundations for the irrevocable delegitimization of slavery colonialism, but at the same time, whose arguments took part in establishing a new form of colonial power - imperial colonialism. The fight between the Jacobins, with Blumenbach at the forefront, and the pro-slavery camp, headed by Christoph Meiners, led to a dichotomy between the temporal and spatial dimensions of human diversity—one of the main epistemological frameworks that ensured the application of Blumenbach's approach by racially minded scholars involved in the legitimization of post-slavery capitalism in the nineteenth century.

The increasing diversity of approaches toward merging the temporal and spatial dimensions of human variation forged the intersectionality of race and class within various racial hierarchies. Reinventing colonial domination through reintroducing racial hierarchies as derivatives of the intersectionality of race and class gave life to intra-racial hierarchies, the division into more and less educable and "civilizable" humans within one race and nation. It served multiple projects of internal colonialism as well as transnational initiatives aimed at establishing surveillance over regions seen as not aligned with the global agenda of security. Reexamining the contribution that those we have come to regard as anti-racist have made to the legitimization of these projects remains an ongoing task.

Echoes of Blumenbach's approach to human perfection, including those defined by freedom from racial prejudice (Squadrito, 1979 112),can still be heard today. The debate between Noam Chomsky and Kay Squadrito in the late 1970s regarding the prerequisites for the human mind to become infiltrated by racism represents one such example (Chomsky, 1975). What makes the human mind more vulnerable to the risk of being saturated by propaganda, either for accepting people as tabula rasa or for relying on rationalist characterizations that justify theories of racial inferiority (Hibbs, 2022)? Does the answer to this question always contain the risk of being transformed in favor of progressiveness and, therefore, the risk of an inevitable mutation in favor of the equilibrium of colonialism? And is there any optimal response to the colonial equilibrium other than critical historicization?

References

Ambridge, L. J. (2012). Imperialism and racial geography in James Henry Breasted's Ancient Times, a history of the early world. *Journal of Egyptian History, 5*(1–2), 12–33.

American Slavery (From the Glasgow Chronicle.) Discussion between Mr. George Thompson and the Rev. R. J. Breckinridge. Fourth Night – Thursday, June 16th. Patriot - Wednesday 13 July 1836 (pp. 265–268).

Archer, M. S. (2003). The private life of the social agent: What difference does it make?" In *Critical Realism* (pp. 29–41). Routledge.

Barbillon, C. (2004). *Les Canons du corps humain dans l'art français du XIXe siècle: L'art et la règle*. Odile Jacob.

Barzun, J. (1965). *Race a Study in Superstition Harper Torchbooks The Academy Library*. Harper Row Publishers.

Baskar, B. (2016). Small national ethnologies and supranational empires: The case of the Habsburg monarchy. In *Everyday culture in Europe* (pp. 65–80). Routledge.

Benedict, R., & Mead, M. (2019). *Race: Science and politics* (p. 38). University of Georgia Press.

Benes, T. (2023). Race, revelation, and the comparative history of religion. In G. Kaplan, & K. M. Vander Schel (Eds.), *Oxford history of Modern German theology* (vol. 1, pp. 1781–1848). Oxford Academic.

Bennett, T., Grossberg, L., & Morris, M. (Eds.). (2013). *New keywords: A revised vocabulary of culture and society*. Wiley.

Bernasconi, R. (2008). A Haitian in Paris: Anténor Firmin as a Philosopher against Racism. *Patterns of Prejudice, 42*(4–5), 365–383.

Biddiss, M. D. (1966). Gobineau and the origins of European Racism. *Race, 7*(3), 255–270.

Bleichmar, D. (2007). Training the naturalist's eye in the eighteenth century: Perfect global visions and local blind spots. In C. Grasseni (Ed), *Skilled visions: Between apprenticeship and standards* (pp. 166–190). Berghahnbooks

Blue, G. (1999). Gobineau on China: Race theory, the 'Yellow Peril,' and the critique of modernity. *Journal of World History, 10*(1), 93–139.

Blumenbach, J. F. (1775). *De Generis Humani Varietate Nativa* [On the natural variety of Mankind] 1775 Göttingane: Typis Frid. Andr. Rosenbuschi

Blumenbach, Johann Friedrich to Daniel Chodowiecki in Berlin Göttingen, 12 January 1782 in Correspondence. In F. W. P. Dougherty (Ed.), *The correspondence of Johann Friedrich Blumenbach* (vol. 1, pp. 1773–1782), *Letters 1 - 230*. Göttingen: Klatt, 2006.

Blumenbach, Johann Friedrich to Johannes Loretz in Berthelsdorf Göttingen, 7 July 1791 in Dougherty, Frank William Peter. *The Correspondence of Johann Friedrich Blumenbach* (vol. 4, pp. 1791–1795). Göttingen: 2012.

Blumenbach, J. F. (1796). *Abbildungen naturhistorischer Gegenstände*. No. 31–60. bey Johann Christian Dieterich.

Böker, W. (2018). Zur Geschichte der Schädelsammlung Johann Friedrich Blumenbachs. *Deutsche Gesellschaft für Geschichte und Theorie der Biologie (Ed.) Annals of the History and Philosophy of Biology, 23*(23), 3.

Böttiger Carl August to Johann Friedrich Blumenbach in Göttingen Weimar, 6 September, 1795. In F. W. P. Dougherty (Ed.), *The correspondence of Johann Friedrich Blumenbach (1791–1795)*, (vol. 4, pp. 645–965), Göttingen: 2012.

Brattain, M. (2007). Race, racism, and antiracism: UNESCO and the politics of presenting science to the postwar public. *The American Historical Review, 112*(5), 1386–1413.

Broca, P., & Blake, C. C. (1864). *On the phenomena of hybridity in the genus homo*. Anthropological society.

Broca, P. (1864). *On the phenomena of hybridity in the genus*. In C. C. Homo Blake (Ed.), Longman, Green, Longman, & Roberts.

Calame, C. (2009). *Poetic and performative memory in Ancient Greece: Heroic reference and ritual gestures in time and space*. Hellenic studies series 18. Center for Hellenic Studies. http://nrs.harvard.edu/urn-3:hul.ebook:CHS_CalameC.Poetic_and_Performative_Memory_in_Ancient_Greece.2009

Chomsky, N. (1975). *Reflections on language*. Pantheon.

Collet, D. (2014). Im Netz der Dinge. Kulturen des gelehrten Sammelns in der Personalunion. In A. Reitemeier (Ed), *Kommunikation und Kulturtransfer im Zeitalter der Personalunion zwischen Großbritannien und Hannover* (pp. 235–259). Göttingen University Press.

Craipain, B. (2024). The laws of indiscipline: Anténor Firmin, racial justice, and the case for a humanist anthropology. *Dix-Neuf, 28*(1), 86–105.

Dagg, J. L., & Derry, J. F. (2024). Patrick Matthew's synthesis of catastrophism and transformism. *Notes and Records, 78*(1), 167–188.

Darlington, C. D. (1961). *Darwin's place in history*. Blackwell, Macmillan.

Davis, D. B. (1987). Reflections on abolitionism and ideological hegemony. *The American Historical Review, 92*(4), 797–812.

Dougherty, F. W. P. (1990). Christoph Meiners und Johann Friedrich Blumenbach im Streit um den Begriff der Menschenrasse. In G. Mann, & F. Dumont (Eds.), *Die Natur des Menschen: Probleme der Physischen Anthropologie und Rassenkunde (1750–1850)* (pp. 89–111). G. Fischer.

Drescher, S. (1990). The ending of the slave trade and the evolution of European scientific racism. *Social Science History, 14*(3), 415–450

Ehler, M. (2003). *Daniel Nikolaus Chodowiecki: "Le petit maître" als großer Illustrator.* Lukas Verlag.

Eigen, S. (2005). Self, race, and species: JF Blumenbach's atlas experiment. *German Quarterly*, 277–298.

Encyclopédie. (1751). ou Dictionnaire raisonné des sciences, des arts et des métiers, volume 1.

Firmin, J.-A. (2000). *The equality of the human races (positivist anthropology)*. Taylor & Francis.

Golf-French, M. (2019). Bourgeois modernity versus the historical aristocracy in Christoph Meiners's political thought. *The Historical Journal, 62* (4), 943–966.

Gordon, L. R. (2018). Chapter 2. Franz Boas in Africana Philosophy. In N. Blackhawk, I. L. Wilner (Eds.), *Indigenous visions: Rediscovering the world of Franz Boas* (pp. 42–60). Yale University Press.

Gould, S. J. (2014). The geometer of race. In *The concept of race in natural and social science* (pp. 1–5). Routledge.

Gray, S. H. (2025). Forster's critique of slavery in Meiners's natural history: On objectivity and cultural creativity. In *Lessing yearbook/jahrbuch LI, 2024* (pp. 157–168). Wallstein Verlag.
Greenberg, M., Chodowiecki, D. (1921). Eighteenth century etcher. *The Brooklyn Museum Quarterly, 8*, 121–131. The Museum.
Hanke, L. (2013). *Bartolomé de las Casas: An interpretation of his life and writings.* Springer.
Harris, K. A. (2019). W.E.B. Du Bois's "Conservation of Races" A Metaphilosophical Text. *Metaphilosophy, 50*(5), 670–687.
Helbig, D., & Nassar, D. (2016). The metaphor of epigenesis: Kant, Blumenbach and Herder. *Studies in History and Philosophy of Science Part A, 58*, 98–107.
Hibbs, D. (2022). The relevance of race in modern philosophy. *Academic Questions, 35*(1). https://doi.org/10.51845/35.1.9
Hünniger, D. (2018). Bilder machen–Charaktere, Stereotype und die Konstruktion menschlicher Varietät bei Johann Friedrich Blumenbach. *Lichtenbergs MenschenBilder: Charaktere und Stereotype in der Göttinger Aufklärung*, 65–77.
Jantz, H. S. (1942). German thought and literature in New England, 1620–1820: A preliminary survey. *The Journal of English and Germanic Philology, 41*(1), 1–45.
Keel, T. D. (2013). Religion, polygenism and the early science of human origins. *History of the Human Sciences, 26*(2), 3–32.
Keel, T. (2018). Blumenbach's race science in the light of Christian supersessionism. In *Johann Friedrich Blumenbach* (pp. 123–141). Routledge.
Keith, A. (1940). Blumenbach's Centenary. *Man, 40*, 82–85.
Kohn, M., & Reddy, K. (2024). Colonialism. In Edward N. Zalta & Uri Nodelman (Eds.), *The stanford encyclopedia of philosophy* (Summer 2024 Edition).
Kontje, T. (2019). The German tradition of the Bildungsroman. *A History of the Bildungsroman*, 10–32.
Kühnast, A. (2020). Johann Friedrich Blumenbach's 'New Hollander'. *Australian Studies Journal Zeitschrift für Australienstudien*, 33/34–2019/2020.
Lawson, B., & Kirkland, F. (1999). *Frederick douglass: A critical reader.* Blackwell.
Lewis, H. S. (2018). Franz boas and friends? Not really. *Reviews in Anthropology, 47*(1–2), 21–38.
Lichtenberg, G. C. (1901). *Chodowiecki und Lichtenberg : Daniel Chodowieci's Monatskupfer zum "Göttinger Taschen Calender" nebst Georg Christoph Lichtenberg's Erklärungen.* Dieterich.
M'charek, A., & Van Oorschot, I. (2019). What about race?. In *The Routledge companion to actor-network theory* (pp. 235–245). Routledge.
Maierhofer, W. (2023). A pro-American perspective on the American Revolution: Johann Heinrich Ramberg's (1763–1840) adaptations of illustrations by Daniel Chodowiecki. *Word & Image, 39*(4), 434–452. https://doi.org/10.1080/02666286.2023.2219594
Mandler, P. (2006). *The English national character: The history of an idea from Edmund Burke to Tony Blair.* Yale University Press
Marino, M. (2021). Naturgeschichte und Rassenklassifikation: Zu Blumenbachs Anthropologie und ihrer Rezeption. In L. Follesa & F. Vercellone (Eds.), *Bilddenken und Morphologie: Interdisziplinäre Studien über Form und Bilder im philosophischen und wissenschaftlichen Denken.* (pp. 73–98). De Gruyter.
Marks, J. (2012). Why be against Darwin? Creationism, racism, and the roots of anthropology†. *American. Journal of Physical Anthropology, 149*, 95–104. https://doi.org/10.1002/ajpa.22163
McLaughlin, P. (1982). Blumenbach und der Bildungstrieb: Zum Verhältnis von epigenetischer Embryologie und typologischem Artbegriff. [Blumenbach and the Formative Drive: On the relation of epigenetic embryology to a typological species concept]. *Medizinhistorisches Journal, 4*, 357–372.
Meiners, C. (1790). Ueber die Natur der Afrikanischen Neger, und die davon abhangende Befreyung, oder Einschränkung der *Schwarzen. Göttingisches historisches Magazin.*
Messling, M. (2016). Text and determination: On Racism in 19th century European philology. *Philological Encounters, 1*(1–4), 79–104.

Michael, J. S. (2021). *The race supremacist anthropology of Christoph Meiners, its origins and reception*. Unpublished manuscript.
Mielke, A. (1988). Hottentots in the aesthetic discussion of eighteenth-century Germany. *Monatshefte, 80*(2), 135–148.
Mudford, P. G. (1968). William Lawrence and the Natural History of Man. *Journal of the History of Ideas, 29*(3), 430–436.
Murison, J. (2011). *The politics of anxiety in Nineteenth-Century American literature*. Cambridge University Press.
Nott, J. C., & Gliddon, G. R. (1854). *Types of mankind: or, Ethnological researches, based upon the ancient monuments, paintings, sculptures, and crania of races, and upon their natural, geographical, philological and Biblical history*. Lippincott, Grambo & Company.
Palti, E. (2005). Romantic philosophy and natural sciences: Blurred boundaries and terminological problems. *Contributions to the History of Concepts, 1*(1), 83–108.
Philalethes. (1864). Peyrerius, and theological criticism. *Anthropological Review*, 109–116.
Prichard, J. (1836). *Researches into the Physical History of Man*. Sherwood, Gilbert & Piper
Rainger, R. (1980). Philanthropy and science in the 1830's: The British and Foreign Aborigines' protection society. *Man*, 702–717.
Reid, D. (2024). British humanitarianism, indigenous rights, and imperial crises: Assessing the membership base of the Aborigines' Protection Society, 1840–73. *Journal of British Studies*, 1–18.
Reill, P. H. (2018). Ethnographic exploration in the Blumenbachian tradition. In *Johann Friedrich Blumenbach* (pp. 177–196). Routledge.
Reus-Smit, C. (2007). International crises of legitimacy. *International Politics, 44*, 157–174.
Richards, R. J. (2000). Kant and Blumenbach on the Bildungstrieb: A historical misunderstanding. *Studies in History and Philosophy of Science Part c: Studies in History and Philosophy of Biological and Biomedical Sciences, 31*(1), 11–32.
Rönnbäck, K. (2012). Enlightenment, scientific exploration and abolitionism: Anders Sparrman's and Carl Bernhard Wadström's Colonial Encounters in Senegal, 1787–1788 and the British Abolitionist Movement. *Slavery & Abolition, 34*(3), 425–445.
Schultz, A., & Wróblewska, M. (2020). Time travel from Berlin to Gdańsk: Daniel Nikolaus Chodowiecki's 18th century journey revisited–contemporary renderings and Interpretations. *Quart. Kwartalnik Instytutu Historii Sztuki Uniwersytetu Wrocławskiego, 2*(56).
Shapin, S. (2019). Why was" custom a second nature" in early modern medicine? *Bulletin of the History of Medicine, 93*(1), 1–26.
Sloan, P. R. (2002). Performing the categories: Eighteenth-century generation theory and the biological roots of Kant's A Priori. *Journal of the History of Philosophy, 40*(2), 229–253.
Sommer, M. (2024). *The diagrammatics of 'race:' Visualizing human relatedness in the history of physical, evolutionary, and genetic anthropology, ca. 1770–2020* (p. 13). Open Book Publishers.
Squadrito, K. (1979). Racism and empiricism *Behaviorism, 7*(1), 105–115.
Stafleu, F. A. The biology of the enlightenment. *Taxon, 16*(5), 431–35. 432.
Steedman, C. (1995). *Strange dislocations: Childhood and the idea of human interiority, 1780–1930*. Harvard University Press.
Topinard, P. (1901). A French anthropologist on Gobineau. *The Open Court, 8*, 11.
Turri, J. (2009). The ontology of epistemic reasons. *Noûs, 43*(3), 490–512.
Wright, G. (1973). New and old views on the economics of Slavery. *The Journal of Economic History, 33*(2), 452–466.
Wright, M. M. (1999). Nigger peasants from France: Missing translations of American Anxieties on Race and the Nation. *Callaloo, 22*(4), 831–52 833.
Wunderlich, F. (2021) Christoph Meiners's Empiricist 'Revision' of Philosophy and Michael Hißmann's Anti-Speculative Materialism. In K. de Boer, & T. Prunea-Bretonnet (Eds.), *The experiential turn in eighteenth-century German philosophy* (pp. 119–138). Routledge.
Yoichi, K., Lee, H. J. S., & Mason, M. (2013). Rule in the name of protection: The Japanese State, the Ainu and the Vocabulary of Colonialism. *The Asia-Pacific Journal, 11*(8), 2.

Zimmerman, A. (2010). Three logics of race: Theory and exception in the transnational history of empire. *New Global Studies,* 4(1).

Chapter 4
The Dissemination and Alteration of Blumenbach's Racial Classification in Nineteenth-Century Polygenetic US-Ethnology and French Anthropology

Malin S. Wilckens

Abstract The widespread circulation and alteration of Johann Friedrich Blumenbach's racial theory across various contexts is a key reason for its differing evaluations. This chapter will explore Blumenbach's practices of "collecting" and comparing skulls, focusing on the methodological changes in relation to concepts of "race" introduced by his successors, Samuel George Morton in the United States and Paul Broca in France. I argue that these changes and adaptations in different places contributed to the stability of Blumenbach's classification. Relationality links "race" and comparing, which is what makes their relationship so revealing. Practices of comparing ensured the stability of "race" through their continuous and flexible actualization in different contexts. The meaning and conceptualization of racial theories evolved through the intricate interaction of various practices, scientific principles, and emerging political and social developments.

Keywords Race · Blumenbach · Physical anthropology · Comparing · Empiricism

At the threshold of the nineteenth century, Johann Friedrich Blumenbach (1752–1840) formulated arguably the most influential racial categories in history, which continue to resonate today. His pivotal role in the research of scientific racism underscores the importance of his work. Blumenbach's theory, however, would not have gained such prominence if it had not been further disseminated, operationalized, and popularized in academic circles. This perspective is crucial for comprehending the enduring influence of Blumenbach's racial theory in diverse social, political, and cultural contexts.

M. S. Wilckens (✉)
Leibniz Institute of European History, Mainz, Germany
e-mail: wilckens@ieg-mainz.de

V. Shmidt and S. De Angelis (eds.), *Blumenbach's Racial Classification*, History, Philosophy and Theory of the Life Sciences 39,
https://doi.org/10.1007/978-3-032-15950-2_4

The production and spread of scientific knowledge were intertwined with that era's societal and political debates and global structures. As this chapter will demonstrate, the local scientific-practical negotiations of a *community of practice* were deeply connected to translocal and global dynamics. This interplay can shed light on the reasons behind the dissemination and alteration of Blumenbach's classification. Scientists' actions are deeply embedded in societal dynamics, as action, practice, and structure are interdependent. Practices embody the social and act as mediators between actions and structures (Reckwitz, 2002; Bourdieu, 2014; Giddens, 1984). Relationality is strongly connected to the presence of others, which is a central guiding mechanism at different levels, proximities, and distances.

An experiential turn in eighteenth-century science affected Blumenbach's convictions about the empirical grounding of his theory of human diversity; it was one of the reasons for his later osteological approach (de Boer & Prunea-Bretonnet, 2021; Bodenmann & Rey, 2018). Combining different epistemological approaches, he was a prominent figure who blurred the lines between rationalism and empiricism. The shift in scientific ideals, virtues, and perceptions of science in the mid-nineteenth century, as identified by Lorraine Daston and Peter Galison (2007), was another significant catalyst for changes in methodology and vice versa and, consequently, for classification and the later dissemination of Blumenbach's theory. Physical anthropology was deeply aligned with this trend, as racial classifications generally resulted from many interacting disciplines and approaches (McMahon, 2018). The idea of an idealized representation of nature, to which Blumenbach adhered, gave way to the understanding of an "objective" representation of a given research object's details. This chapter will demonstrate the profound significance of this shift, which was not least prompted by racial science and is shown in the alteration of his theory, even if a practice-specific view reveals a more nuanced picture. Practices of "collecting" and comparing, which Blumenbach, as one of the first anatomists, combined perfectly, were affected by discussions of morphology, whose focus changed by the mid-nineteenth century from the idea of an ideal type to the sum of individual traits. This chapter shows how different multilayered practices have produced a specific understanding of "objectivity" in the natural sciences. As skull "collecting" gained popularity, making skulls even more of a commodity, comparisons turned from the descriptive features of an individual, gaining ideal types, to measurable practices of entire skull series. Whereas Blumenbach's ideas of empiricism were still in line with idealizing approaches, following anthropologists connected them with "objective" statistical data.

The argument put forth in this chapter is that it was precisely the alteration and adaptation of Blumenbach's theory in different contexts that made his classification so stable. The same can be said about the practices of knowledge production, e.g., criteria to define "races" by ordering and differentiating racialized groups. Therefore, a further thesis posits that practices of comparing skulls, central to the racial classifications made by comparative anatomists, were given such stability through their constantly changed actualization in different contexts (Wilckens & Gärtner, 2022).[1]

Practices of comparing are characterized by a triadic structure in that they relate at least two objects of comparison (*comparata*) in a certain respect (*tertium comparationis*). Comparisons are always situated in specific contexts. In addition, every comparison is based on an assumption of similarity, without which no comparative operation would be possible. Although comparisons radiate methodological neutrality, they are hardly neutral, as actors select the objects of comparison and *tertia* and thus direct the comparative gaze (Epple et al., 2020; Candea, 2018; for critics of the comparative method see Boas, 1896; Radhakrishnan, 2009). Comparing is inscribed in the order of people as it negotiates the relationship of similarity (being human) and difference. "Race" and comparing are based on this relationality, which makes their relationship so revealing. Comparing also enables the connection of different contexts and localities, as *comparata* and *tertia* can be dis- and re-embedded (Epple, 2015; Giddens, 1990), making racial classifications adaptable to various purposes and contexts.

The dissemination and alteration of Blumenbach's racialized theory, in which he ordered the world into four, and, later, five "races"—American, Caucasian, Ethiopian, Malay, and Mongolian—will be traced through the American School of Ethnology, notably through the contributions of Samuel George Morton (1799–1851) and the Anthropological Society in Paris, in particular, through the work of Paul Pierre Broca (1824–1880). These researchers played a significant role in shaping and adapting Blumenbach's theory, contributing to its stability and endurance. By focusing on practices of comparing, it will be shown how measuring became a key congruent to making "race" more "objective." The meaning and theorizing of "race" shifted through the complex interplay of different practices, scientific ideals, and new thematic, political, and societal developments.

An important dimension unites these three researchers in the field of racialized sciences. They were all empiricists who not only worked with haptic "material," but also "collected" skulls for their research purposes, albeit in different ways. In doing so, they followed an Enlightenment ideal of empirical and observation-based science (for different paths in Enlightened thinking see Conrad, 2012; Winterer, 2016). Morton became one of the most important comparative anatomists, following in Blumenbach's footsteps, although shifting his classification to polygenesis. While advocates of monogenesis assume a common origin for all humans, polygenesis asserts that humans have different origins. The polygenist Broca based his research heavily on the American School of Ethnology, especially its most prominent representative, Morton.

The chapter first explains why Blumenbach's practices and theory gained popularity, which structures shaped his research, and what made his approach transferable. It then examines Morton's research practices and situates them in the context of the

[1] This argument is inspired by the Collaborative Research Center 1288 "Practices of Comparing" at Bielefeld University.

removal of Indigenous peoples and slavery. The French anthropologist Paul Broca will be analyzed in the context of the formation of the French Republic. In this way, it can be shown how and why racial theories in general and Blumenbach's classification in particular were disseminated, popularized, and altered.

4.1 Why Blumenbach? Enlighted and Colonial Practices in an "Ever-Expanding" World

Blumenbach's research covered various topics related to natural history during the transition from the eighteenth to the nineteenth century (Rupke & Lauer, 2018). However, he became most famous for his racial classification based on skulls. He was by no means the first and only person to categorize people into different varieties, but his typology in particular influences science until today. What then distinguished Blumenbach from his colleagues and predecessors? What caused his classification to be reproduced and transformed across diverse contexts for generations of scholars who followed? I argue for an answer grounded in three main reasons: his unique skull collection, his symmetrical fivefold classification, and its naming.

Blumenbach mainly relied on the classification of Carl von Linné and Georges-Louis Leclerc de Buffon, who categorized humans based on travelogues and illustrations, with skin color being the dominant differentiation marker. Initially, he adopted Linné's fourfold continental scheme in ordering the human world. As James Cook's world travels brought new insights into cartography, Blumenbach added a fifth human variety—even before he began extensively "collecting" skulls. He also shared research interests with his colleague and conservative advocate, Christoph Meiners. Meiners based his human classification on common visual comparison methods and travelogues. For him, there were primarily only two varieties: beautiful and ugly. He used the designations Mongols and Caucasians, which, despite their diverse valuation of human "races," later appeared in Blumenbach's work (Meiners, 1785; Golf-French, 2023; for Meiners' connection of natural and juridical inequality and the term "juridical racism" see Hüning, 2023; for the dispute with Blumenbach see Dougherty, 1996). While Meiners criticized Blumenbach for determining entire "races" based on one "object," namely, the skull, they shared the semiotics of "race." However, Blumenbach's complete, earth-encompassing classification and empirical approach through skulls made him the founder, especially of the Caucasian variety.

With Dutch physician Peter Camper, Blumenbach shared the osteological and aesthetical approach (Painter, 2010). It is not least because of him that Blumenbach made "white" the original color of the human "race." Camper had already dealt extensively with classifying people based on skulls and utilized the *facial angle* to determine a scale from apes to the Greek god Apollo as the ideal image of human beings. Blumenbach considered the *facial angle* insufficiently complex and used the *norma verticalis* as a classifying method by placing the skulls horizontally on a table and viewing them from above and behind. Camper's student, Samuel Thomas von Soemmerring, also focused on human differentiation, especially between Black and White people. He primarily dissected brains and skull bones and believed he could make intellectual judgements based on measurements of cranial capacity. The Black-White dichotomy[2] was a prevalent subject in medical research, as well as in ordering humankind, not least because of the transatlantic slave trade (Malcolmson, 2013; Mazzolini, 2014; Parfitt, 2020). This topic also played a significant role in a dispute between Blumenbach and Soemmerring concerning the intellectual abilities of Black people. Often, in his writings and letters, Blumenbach emphasized the fine intellectual abilities of Black people, in contrast to Meiners and Soemmerring, who tried to legitimize the degraded status of Black people (Dougherty, 1985).

The interplay of the different arguments proposed by his colleagues was reflected in Blumenbach's theory, as the focus on practices of comparing and ordering demonstrated. Nevertheless, Blumenbach's work differed from his colleagues' approaches. While the late Enlightenment affected them differently, with empiricism becoming prominent as a research ideal, only Blumenbach succeeded in amassing a considerable collection of human skulls (for his collection's evolution see Böker, 2018). It was, therefore, the haptic research "object," and in significant numbers, which ensured such popularization of his classification.

On the threshold of the nineteenth century, in the comparison arrangement, the main *tertium* skin color changed to a new "object" —the skull—which was already recognizable with Camper and Soemmerring. The skull became so prominent because, in contrast to skin color, it could be "collected," measured, and compared in various ways (Wilckens, 2022a). These practice-orientated and epistemological explanations were related to the general assumption that the skull is the center of the soul, spirit, and intellect. Blumenbach amassed many skulls through his scientific network, stretching all continents. Joseph Banks, one of his central intermediaries in gaining skulls, sent him skulls from the South Sea, Australia, and the Caribbean. As botanist of Kew Gardens and president of the Royal Society in London, he was closely connected to political officials. He, therefore, profited extensively from the British colonial structures for his scientific endeavors—as did Blumenbach. In 1790, he wrote to Blumenbach:

> Since I had the honor of sending you the Carrib skull to which you have done so much honor in your Elegant publication I have not been so successfull as I could have wishd[.] the Loss of the Bounty which was bringing home bread fruit from the East to the West Indies & of the Guardian destind for the Releif of Botany bay have together deprivd me of the Fruits of requests on that subject made to both their Commanders[.] (Banks to Blumenbach 1790)

The Bounty was a British Admiralty ship that set off on a voyage to the South Seas in 1787 under the command of Captain William Bligh, to bring the breadfruit tree from Tahiti to the Antilles in the Caribbean. Banks commissioned Captain Bligh to bring back skulls for Blumenbach. However, due to the mutiny, the ship did not return to England, and Blumenbach did not receive the expected skulls. Skulls did not always reach Göttingen, influencing ordering practices. Skulls from Asian territories reached Blumenbach via Georg Thomas von Asch, a former medical student in Göttingen

[2] The terminology refers to the accepted spelling. It covers, of course, different groups of people.

and later State Counselor in the Russian Empire. His close connection to Göttingen, as well as to anatomical centers, enabled a variety of skulls of different origins to be sent to the Göttingen professor, thereby profiting from the internal colonialism of the Russian Empire. The theologian Johannes Loretz "collected" skulls from missions in North America. A missionary of the Moravian Brotherhood, Jesper Brodersen, wrote to Loretz from Greenland in 1792:

> There is, however, a difficulty in this, which does us no particular honor, but which can be excused to some extent in the case of a nation such as Greenland, namely that we must try to get hold of it quite secretly from the Greenlanders, since the living would not forgive us so easily if they realized that we were sending the skulls of their deceased compatriots across the sea wrapped up. However, it will be possible; I have advised my dear brothers in all three places to consider this when the opportunity arises, and I will not neglect it. (translated by author; Brodersen to Loretz 1792)

Most of the grave openings had to be carried out in secret, as "procurers" were aware of the illegitimacy of their actions. All of them profited from colonial or inner-colonial structures to procure skulls. Interestingly, Blumenbach never left Göttingen himself to "collect" skulls. He only used his pen and a letter, presented the purpose of his research, and offered books or other courtesies in return (cf. Blumenbach to Banks 1794a, 1794b; Blumenbach to van Geuns 1795). Numerous skulls were acquired during wars, from colonial sites, stumbled upon accidentally while travelling, or obtained through post-mortem examinations in anatomy departments. By the end of his life, his collection had counted more than 240 skulls (Böker, 2018). The creation of his unique skull collection was closely linked to colonial and Enlightenment structures. Colonial contexts, in particular, were decisive enabling conditions that allowed him to procure skulls from almost all regions of the world and thus fulfilled the Enlightenment-inspired urge for objects worldwide.

While "collecting" had a global dimension, comparing and categorizing these skulls was done in Göttingen with an imaginary global dimension. Analogy and zoological critique were the central methods underlying Blumenbach's monogenetic argumentation. He investigated the relationships between specific physical characteristics within a species compared to the ratios of these characteristics in another species, deducing the same causalities of differences in tame animals as in humans (Blumenbach, 1789; Lauer, 2014).

For his typology, he used comparative practices. He compared descriptively the shape of the forehead, skull, jaw, eye socket, nose, etc. The manifold small-scale differences between each skull made defining varieties difficult. It was, therefore, necessary to construct ideal types, which he reached through European aesthetics of symmetry, being his central *tertium* (*umbrella-tertium*, includes diverse *subtertia*) (Blumenbach, 1795, 1798).[3] Blumenbach visualized what he considered to be the three most ideal skull shapes with the help of the *norma verticalis*, focusing on the

---

[3] In contrast to Michael (2021), who pleads to downgrade aesthetics within Blumenbach's typology, I argue (especially because he used positive aesthetic ascriptions also for other varieties, not only for the Caucasian skull) it is the most central for his typology. Focusing on practices of comparing, symmetry, and, therefore, culturally determined aesthetics are omnipresent, without excluding other aspects, like the condition of skull, etc.

total habitus of the skull. This arrangement enabled him to strengthen the thesis of an original skull type and to introduce a symmetrical aesthetic into his overall typology. These idealizing comparisons corresponded to the standard scientific practice of the time (Boulter, 2021; Daston & Galison, 2007). He also relied on old, well-established practices by comparing the "objects" with portraits and images to value, on the one hand, the correctness of images and, on the other, the collector's data, which strengthened his empirical evidence-based focus. Only at the end were the anatomical descriptions given geographical classifications, which spatialized the varieties.[4] Even though he amassed many skulls, he grounded his typology only on a few heads.

The center of his fivefold classification was the Caucasian variety. The Georgian, Caucasian skull, representing the White "race," was for Blumenbach "in general, according to our judgment of symmetry, the most beautiful and well-formed facial shape" (translation, Blumenbach, 1795: 177f.) (on the problematic of translation see Michael, 2017).[5] For Blumenbach, the skull of the Georgian woman was not only the most beautiful, it was also the reference point against which he compared and evaluated other skulls. According to him, humans degenerated in two directions (for the meaning of degeneration in Blumenbach's work see Junker, 2018), which can be recognized as a temporalization, in that Blumenbach assumed a distance from God's creation. By introducing a historicity of nature, he not only turned away from one comprehensive moment of creation but also introduced insightful results for life sciences by combining paleontological-inspired life forms with the development of humankind (Zammito, 2018; on the historicization of anthropology see De Angelis, 2021). He referred to one "extreme" as the Ethiopian type and the other as the Mongolian, identifying the American and Malay varieties as transitional forms. Despite these fluid transitions, they would each display strong national characters. Blumenbach was barely able to resolve this contradiction. In 1795, he wrote, "[I]t is therefore not surprising that a merely arbitrary classification of these varieties can occur" (translation, Blumenbach, 1795: 285).[6] The emphasis on imperceptible transitions confirms the difficulty of finding suitable *tertia* that could serve as criteria for difference (Epple, 2020). The concrete marking of difference and the unambiguous naming posed a substantial challenge for natural scientists, which can also be seen in the contingency of taxonomic classifications in general.

---

[4] The allocations of human groups varied over time. Ultimately, the Caucasian variety included all Europeans, Western Asians, and North Africans, except the Sámi and Finns. Blumenbach counted all eastern Asians, the Sámi and Finns of northern Europe, and the northernmost Americans, the Inuit, in the Mongolian variety. All other inhabitants of the Americas counted to the American variety. The Ethiopian variety included all Africans except the northernmost. The Malay variety consisted mainly of the South Sea Islanders.

[5] According to Blumenbach, other skulls could also be well-formed, symmetrical or elegant, especially if they corresponded to the respective ideal type of their variety. However, the positive attributions occurred more frequently when Blumenbach presented skulls individually. When it came to the superordinate ideal types of the two "extremes" (Mongolian and Ethiopian variety), negative connotations predominated.

[6] When it wasn't about the skull, the face, or skin color, Blumenbach detached himself from the scheme of five types and primarily described individual nations.

The second edition of Blumenbach's work on humankind had named the first variety *Greek-Egyptian*. Why had it been relocated to the Caucasus in his final edition? He mixed physiological, aesthetic, and religious arguments. For Blumenbach, science and religion were not contradictory; he referred to divine creation in many passages of his works. Christian intellectuality had a strong influence on the natural history of humans, as Keel has analyzed in depth (2018a, 2018b; Gerbner, 2019). Even if Blumenbach did not explicitly refer to the biblical story of Noah (though he did refer to a mountain), it certainly played a role in his argumentation. Much more important, however, was the assumption of many, mainly French travelers, that the most beautiful people, especially women, lived in the Caucasus. According to Blumenbach, God, as the perfect creator, must have created the most beautiful people. Conversely, this means that where the most beautiful people live must be the origin of their species.

Blumenbach stated that the beautiful people from the Caucasus are not only found in ancient statues, such as Venus Medici, but the skull of the Georgian woman also reifies the bust of Clytie (Klatt, 2008). The reference to Greek antiquity was a necessary legitimization strategy to relocate the first variety to the Caucasus. Blumenbach could only take this step by proving their similarity. Additionally, Meiners' designation played a crucial role in naming. Most important, though, was obtaining an intact Georgian skull in 1793 (ibid.). He aimed to visualize his types, needing representative skulls. This "accidental find" (von Asch to Blumenbach, 1793), as von Asch has called it, played a significant role in Blumenbach's typology and showed the interplay and contingency of "collecting," comparing, and semiotics.

The discursive negotiation and naming practices reveal an inconsistency in Blumenbach's terminology. The Göttingen professor rarely used his various terms consistently. He did not clearly define "race" until 1797, adopting it from Immanuel Kant (Klatt, 2010; for a link between Kant and Blumenbach in the life sciences see Sloan, 2002). For Blumenbach, the difference between variety/species and "race" was ultimately the necessary inheritance of a characteristic among "races." This late public adoption of the difference and precise definition of the concept of "race" shows, on the one hand, the contested and variable terminology and, on the other hand, the epistemological uncertainty in Blumenbach's taxonomy and theory.

With his symmetrical fivefold classification, he created, on the one hand, a holistic and, on the other, a reductionistic order of people that could be adapted and modeled in different contexts. Blumenbach always emphasized the fluid transitions between "collected" skulls and the impossibility of a clear distinction; yet, he created a classification. His empirical findings were, therefore, inconsistent with his final proclaimed categorization and left empty spaces which could be easily filled by the next generations of those who aimed at hierarchizing humanity through racial division. This argument is further supported by Blumenbach organizing people into his note system in 61 different fascicles—a much more differentiated classification system (Cod. Ms. Blumenbach V, SUB). The Enlightenment ideal of empirical and categorical research led to other methodological regimes of evidence, additionally making it difficult for him to render "degeneration" and fluidity visible, leading to a contradictory epistemology. One argument is that Blumenbach created a classification because he sought and needed to inscribe himself in the socially shared Enlightenment practices of ordering, comparing, and categorizing in order to be heard. Blumenbach thus positioned his research directly in his time's major global debates and new scientific ideals. Because "collecting" brings a very ambiguous picture, classification and designation require straightforwardness.

Blumenbach's knowledge practices can be understood at the threshold of modernity, as he combined old methods of analyzing travelogues and comparisons of images with empirical skull comparisons (Hünniger, 2018). Even though Blumenbach vehemently advocated the unity of the human species, he introduced an aesthetic hierarchy and a European standard of comparison into his "racial" typology. The supposedly neutral comparison method shows deep entanglements with a colonial epistemology of knowledge, making his classification attractive in various contexts. He could not set aside his ideals of beauty, interwoven with Greek aesthetic features, which contributed to an epistemic hierarchy (Painter, 2010). Although Blumenbach used his research to argue against slavery by proclaiming physical and intellectual unity, the clear categorical distinction between humans allowed for strong argumentative lines of demarcation and made his theory so applicable to racial theories of polygenism.[7] What Blumenbach could not know was the scope of the adaptation and alterations of his theory in the following decades, showing the importance of unintended consequences in natural science.

4.2 Morton—Imperial Practices, the "New World," and the Origins of Humankind

Samuel George Morton, a physician and anatomist from Philadelphia, born into a Quaker family, enjoyed a high reputation throughout his life, above all, for his amassed collection of human and animal skulls. His contemporaries highly praised him: "No Man in the United States, in fact, not the bravest soldier or the most illustrious statesman, was better known or enjoyed a wider and more enviable reputation abroad" (Smith, SGMP, APS). He has long retained his status as the leading figure in the so-called American School of Ethnology/Anthropology,[8] which, at the beginning of the nineteenth century, was primarily concerned with the origins of humankind. Morton began to study medicine in Philadelphia and continued his studies at the famous medical school in Edinburgh. These transatlantic ties continued during his academic career, as he stood in close exchange with French and British anatomists in particular and tried to popularize his ethnological works in Europe (Poskett, 2015). The so-called *res publica literaria* survived its European heyday well into the nineteenth century among American intellectuals, including Morton (Winterer, 2012). The lively scientific exchange, the collecting of natural history objects, and the

[7] Not to be overlooked is the fact that monogenetic theories can be equally racist.

[8] Most commonly referred to as the American School of Anthropology, here, I use Ethnology, as Morton more often used this designation for the study of human "races."

founding of many scientific journals can be summarized under the term "American Enlightenments," the specific nature of which, in addition to the solid Protestant character, was expressed above all through plantation slavery and contact with Native Americans (Winterer, 2016). These themes and imperial structures deeply guided Morton's anatomical studies. Blumenbach was both the starting point and a reference for the legitimization of his research.

Just a few decades after Blumenbach's death, it became clear how famous his theory would become, and all subsequent craniologists endeavored to follow in his footsteps: "But if the celebrated German has received meet praises for this work, shall we not also claim for an American physician and naturalist a share of applause for labors more assiduous, and results far more considerable and valuable?" (Meigs, 1851: 22). In what respect were Morton's findings considered more valuable and scientifically sound? Where did he differ from Blumenbach? As will be demonstrated, Morton's research practices were deeply intertwined with imperial politics and adhered to a distinct epistemological framework.

References to the father of craniological "race" theory, Blumenbach, could be found almost everywhere in the obituaries and memories of Morton's death (for a critical review of Morton's memoirs see Michael, 2020). At the same time, they illustrated Morton's scientific concerns:

> The object proposed primarily being the determination of ethnic resemblances and discrepancies by a comparison of crania, (thus, perfecting what Blumenbach had left lamentably incomplete,) the work could not be commenced until the objects for comparison were brought together. (Patterson, 1854: xxviii)

It is not only the reference to the German, who was considered to have too few "objects," that becomes apparent, but also Morton's adopted global "collecting" practices to obtain the *comparata* to pursue skull comparisons. This was closely related to his earlier interest in fossils, such as shells, to study the geological age of the "New World," promoting paleontology (Winterer, 2024). Like Blumenbach, Morton received skulls from cemeteries, war sites, archaeological burial grounds, or imperial contexts by approaching colleagues, especially military doctors and political actors at home and abroad. In 1832, Morton asked his colleague William Ruschenberger for skulls: "Can you not procure me some skulls of native Indians—or ancient Peruvians—or South Sea Islanders—any thing in the shape of a skull will be a treat to me" (Morton to Ruschenberger, 1832). His focus on Native Americans highlights his research interest in the origin of humans and shows deep entanglements between epistemology and context. William Wood, commissioned by Morton's colleague, Daniel Drake, had been travelling for a few days to get Morton skulls from ancient burial mounds: "Agreeable to request, I examined many of the aboriginal Mounds and burying-grounds, in the vicinity of Big-bone Lick, for Crania, with but partial success" (Wood to Drake, 1833). The skulls were often fractured or crumbled on the surface. Wood described the difficulties:

> We dug over many rods before we could obtain a bone, sufficiently dense to bear removal from its earthly bed. From two to four hands were employed, nearly three days, in digging and examining the places above named, before I succeeded in procuring a simple specimen. (ibid.)

It turns out that skull "collecting" involved many people and was a tenacious and lengthy endeavor. In addition, it was not easy to find the graves and to open them unnoticed, as the ornithologist John Kirk Townsend wrote in 1835:

> It is rather a perilous business to procure indian's sculls in this Country – The natives are so jealous of you that they watch you very closely while you are wandering near their mausoleums & instant & sanguinary vengeance would fall upon the luckless wight who should presume to interfere with the sacred relics. (Townsend to Morton, 1835)

So-called "Indian" graves were sacred burial sites that embodied deep connections to Indigenous ancestors (on the changing value of human remains see Wilckens & Kurzwelly, 2022). Like Blumenbach, Morton did not usually procure skulls himself. Both skull "collecting" efforts were characterized by a high level of knowledge transfer, object transfer practices, and reciprocity with their procurers. Unlike Blumenbach, however, Morton also exchanged skulls, as many more people were now involved in "collecting." Additionally, he bought many skulls, making them more of a commodity (Fabian, 2010). While Blumenbach still had to vindicate the significance of comparing skulls, they were the unquestioned departure point in Morton's works. As for Blumenbach, Morton repeatedly reflected on comparing being the central practice for an epistemology of "race": "As our means of comparing the races of men become more extended, our classifications will of course improve; and meanwhile we must rest content with an approximation to accuracy" (Morton, 1839: 4).

In his most famous work, *Crania Americana* (1839), he referenced Blumenbach right at the beginning but shifted the theory to the spectrum of polygenesis and hierarchized the "races." He asked "whether it is not more consistent with the known government of the universe to suppose, that the same Omnipotence that created man, would adapt him at once to the physical, as well as the moral circumstances in which he was to dwell upon earth?" (Morton, 1839: 2f.). This question not only demonstrated how creationism turned to polygenism with the persistent inclusion of religious explanations, most notably the primacy of the Caucasian "race" (Keel, 2013). It also showed that another characteristic had found its way back into the theory: morality and people's intellectual abilities. Blumenbach had always refused to draw this distinction. Morton adopted Blumenbach's five-fold classification and the respective designations, without mentioning Blumenbach's doubts about the straightforwardness of this typology. Blumenbach's classification was incomplete, "yet it is, perhaps, the most complete that has hitherto been attempted" (Morton, 1839: 5). Therefore, Morton "collected" far more *comparata* and subdivided the five "races" into twenty-two families.

In *Crania Americana*, he aimed to discover whether Indigenous Americans belonged to one "race" and whether the Indigenous people living at his time had the same origins as those living long before. He paid particular attention to the many burial mounds along the Mississippi Valley. The mounds had two functions: on the one hand, they were supposed to prove the possibility of a "civilized" nation on the American continent, a point that authors like Cornelis de Pauw always denied, and on the other hand, they were supposed to show that the contemporary "Indians" were not the builders of these graves. Morton thus argued that, first, it was quite possible to build a modern civilization in America and that, second, the removal of the Indigenous people was justified, although implicitly, as they in no way resembled a "modern civilization." Thus, Morton reinforced the *mound builder myth* popularized then, strengthening the thesis of a *lost ("white") race* (Colavito, 2020; Kassabaum, 2021; Mann, 2003). His main aim was to portray Indigenous people as a separate "race," entirely distinct from Caucasian people (Morton, 1839, 1844a). To support his mound builder thesis, he divided the Americans into two families: the Toltecs, who were considered the predecessors of the Aztecs and the builders of these mounds, and the Americans, with the former being intellectually superior.[9] Driven by the presence of Indigenous people and the debates over their removal, his research can be interpreted as a scientific legitimization of the westward expansion of the United States.

Crania Americana was not just about Indigenous people; Morton also wanted to bring Blumenbach's classification into a hierarchical order, therefore, combining comparison with a practice Blumenbach did not apply, namely, measurement. He not only referred to Camper's *facial angle* but also to Blumenbach's *norma verticalis*, leaving it up to the reader to apply. He was convinced of its function, but it could not be measured. Morton constructed new apparatuses like the facial goniometer and measured the skulls in twelve ways, introducing a new central *tertium*: the skull's capacity.[10] He concluded that Caucasians had the largest brain volume, followed by Mongolians and Malays, with Americans and Ethiopians in the bottom two positions.[11] By selecting his samples, he excluded some of the small Hindu skulls. Interestingly, he did not reflect these issues in the other samples, which showed a bias in producing the sample-*comparata* (for a discussion of the correctness of his measurements see Gould, 1978, 1981; Michael, 1988; Lewis et al., 2011; Weisberg, 2014). Morton chose the skull's capacity as central *tertium* because he assumed a correlation between brain volume and intellect[12]:

> Some benevolent minds deny the moral & intellectual differences observable in the different races of men. This conclusion is the result of impulse or feeling, but not of observation or study. We see these differences in families of the same race & in individuals of the same family. In the same domestic circle some are wise & some are fools. What cannot be denied on the small scale exists also on the great. (Morton, 1842, LCP)

Drawing this connection, he followed a widespread assumption about the correlation between brain size and intelligence, closely connected to the new popular

---

[9] He divided the American family into four further branches: Appalachian, Brazilian, Patagonian, and Fuegian.

[10] Morton took up measuring practices of Soemmerring and Tiedemann and identified four different cranial chambers. He already surmised that the anterior one was more important.

[11] In his final measurement table, the American "race" was at the bottom.

[12] I reject arguments that claim Morton saw no relationship between cranial volume and intellect or character, and those that ascribe an anti-racist viewpoint to him (Collins Cook, 2006; Stanton, 1960).

science of phrenology.[13] His German colleague Friedrich Tiedemann was of the same conviction, measuring the brain volumes of diverse "races." However, his data showed no difference in brain size between Black and White people, which indicates that Morton's interpretations were far from inevitable (Morton's book could also be interpreted as a reaction to Tiedemann; see Wolff Mitchell & Michael, 2019). By the mid-century, Morton followed a new epistemological turn in gaining somewhat objective data. Through his final numeric ranked table, he simplified the measurements by focusing solely on the average brain capacity. The numerical data should not obscure the fact that Morton's measurements were preceded by an almost one-hundred-page essay discussing the cultural practices and customs, character traits, architecture, skin color, and body structure of different human families, to support his statistics and vice versa. Here, Morton's research touched on the anatomical, archaeological, and even philological fields of ethnography. In addition to Blumenbach, Morton primarily referred to the Europeans Carl von Linné, Georges-Louis Leclerc de Buffon, Georges Cuvier, Julien-Joseph Virey, and Jean Baptiste Bory de Saint-Vincent. Morton's reception clearly shows how strongly the French natural history of "man" had influenced him. However, when it came to American skulls, he primarily referred to scientists from the American continent, especially to his skull suppliers.

What he adopted from the German professor on a far larger scale was the visualization of the entire skull collection (on the importance of seeing and visualization in Morton's work see Armstrong-Fumero, 2014; Sommer, 2024). More than 70 lithographs were the main attraction of his work, which was intended to appeal to a wide readership and visually support his thesis. He also owned a drawing of a Botocudo skull from Blumenbach's collection, which Maximilian Prinz zu Wied-Neuwied sent to Morton (Wied-Neuwied to Morton, 1837). Blumenbach only portrayed the five central pattern skulls and some others in his *Decas craniorum*. Nevertheless, sensual perceptions guided his entire typology, whereas Morton's was based on supposedly objective figures due to changing epistemological platforms. However, Morton did not manage entirely without sensory impressions. Aesthetics in science were at a threshold.

Morton's interests extended beyond the American "race." He was concerned with the overall order of the world and its hierarchization, as well as with slavery. Egyptian skulls formed the basis of his next book, which the English Consul and later fellow of the American School of Ethnology, George Gliddon, had almost exclusively "collected" in Egypt. As Morton explained to Gliddon: "[I]n order more fully to make my comparisons, it is very important that I should get a few *heads* of Egyptian mummies from Thebes, &c." (Morton to Gliddon, 1837). Gliddon managed his request very well and robbed 143 skulls for Morton in about two years, with great enthusiasm:

[13] Whether Morton was an advocate of phrenology is not definitively clear. The phrenologist George Combe presented a phrenological table based on Morton's skull collection in *Crania Americana*, which suggests that Morton was quite convinced by its claims, yet strategically refrained from applying the science himself, as it was harshly criticized from the beginning for being unscientific.

> Indeed, it has afforded me a sort of rascally pleasure; and I could make you laugh at my numerous expedients in the resurrection line. Many a chuckle have I had with my Snake Hunter (one of the disciples of the Psilli School) at the success of our ruses spite of the vigilance of Priests, [?], Guards and Families, to abstract skulls from Convents, Tombs, Sanctuaries and mummy Pits; especially when we required to keep our operations secret. (Gliddon to Morton, 1839)

Morton studied the ancient Egyptian civilization in detail for *Crania Aegyptiaca.* Interestingly, this book also began with Blumenbach, who examined some mummies in London while visiting Joseph Banks. It seems that every serious craniological study had to rely on Blumenbach. Morton wanted to determine to which "race" the ancient Egyptians belonged, thus, how their relative position could be determined, and whether the civilization ascended or descended the Nile; in other words, whether it had its origins in Egypt or Ethiopia. He followed Blumenbach's call to examine old monuments in their temporality with regard to skulls. Morton's research documents contained numerous notes, drawings and excerpts about ancient Egypt. He meticulously recorded all noteworthy details about their origins, their architectural achievements, such as the pyramids and the Sphinx, their physical characteristics, and their connections to Ethiopian, Greek, and Roman cultures, drawing from sources such as Blumenbach, Plato, Strabo, Herodotus, Prichard, Wilkinson, Pettigrew, or his rival Tiedemann, as well as various French authors. Everything revolved around the primordial nature of racial diversity. Like Blumenbach, Morton refused the presumption that the Egyptians belonged to the Mongolian "race" and affirmed Blumenbach's assumption regarding ethnic diversity in old Egypt. By analyzing the skull shapes and capacities from skulls of ancient pyramids, he osteologically determined an alleged Caucasian predominance in ancient Egypt, much like his German fellow. He concluded that the Caucasian "'race" consists of three skull types: the Pelasgic, the Semitic, and the Egyptian. Morton deduced two main arguments: the position of the Black "race" was the same then as it is today, "that of servants and slaves" and that "[t]he physical or organic characters which distinguish the several races of men, are as old as the oldest records of our species" (Morton, 1844b: 66). The human relics from the bygone era of the "monumental Egyptians" were intended to substantiate the theory of polygenesis.

Morton's variable use of the concept of "race" became evident in his labeling practices. Later in his work, he replaced "race" as an overarching category, adopted from Blumenbach. He defined various subgroups of the Caucasian variety as "races" and replaced the overall designation by group or family. This was also because, by the end of his studies, he was increasingly convinced that more than five primordial "races" existed, reinforcing his polygenetic argumentation. He further strengthened his polygenist argument by theorizing the mixing of "races." By introducing four degrees of hybridity, he rejected Buffon's species concept, which recognized fertile offspring as the species' criterion (Bitter-Smirnov, 2022; Morton, 1847). According to his final measurement table, the Teutonic, German "race," including Anglo-Saxons, Anglo-Americans, and Anglo-Irish, would possess a larger brain than any other people, thereby hierarchizing "races" within the Caucasian group. A stronger national differentiation was recognizable towards the end of his lifetime, except for African people, who, for him, except the Egyptians, were an almost homogeneous group. The ever-expanding new US-American nation was a further driving force in his changing practices.

The epistemological notion of an ideal type of human being gave way to the assumption of quantitative averages. The practice of quantitative measurement and the more significant number of *comparata* were thus decisive changes that made it possible to place the classification on a supposedly more objective basis. Transparency, verifiability, and falsifiability became epistemologically fundamental premises of natural sciences. The change in *tertium* from symmetry to cranial capacity was another critical development. Morton adopted the "collecting," classification, and naming practices from Blumenbach but modified some of them. Significantly, he added practices of measurement, which quantified Blumenbach's initial classification and hierarchized it according to intellectual ability. Different political contexts and different scientific understandings were the driving forces behind these alterations. Morton's theory fit directly into his country's current, highly explosive political debates, making it applicable in various societal contexts (Fredrickson, 1971). His research not only profited from colonial and imperial structures but reinforced them on an epistemic dimension by promoting epistemic violence.

4.3 Broca—Nation Building, Pre-History and Craniometric Practices

By the mid-nineteenth century, a new centrum of craniological research emerged. Its leading biological anthropologist was Paul Broca, today known for his findings in neurological linguistics, especially the *Broca-Areal*. What is relatively unknown is that he followed in the footsteps of his predecessors Blumenbach and Morton, popularizing craniological racial theory. In 1859, he founded the Anthropological Society of Paris and began to amass a vast collection of human skulls and compare them to theorize "race" and human origin (Blanckaert, 2009). Even though Darwin's *On the Origin of Species* emerged in the same year, the question of the dissemination of humans was far from being consensually solved. Broca was well aware of Darwin's theory but rejected it as unverifiable: "Now, is Mr Darwin right or wrong? I don't know, I don't even want to know. I find, in the things accessible to science, sufficient nourishment for my curiosity, without losing myself in the night of origins" (translation, Broca, 1862a: 314).[14] Broca criticized Darwin for speculation and lack of experience. To create a scientifically sound theory, he followed Blumenbach with an empirical approach, relying on crossbreeds to create a valid theory concerning the hybridity of "race" and, foremost, "collected" skulls to order and hierarchize humans—both disseminating the theory of polygenesis.

[14] It should be added that although Broca himself dealt with the various origins of human beings, he did not specify where and when they began, only that they were multiple.

This section will demonstrate that Broca's racial theories were deeply embedded in French nation-building. Right after the Anthropological Society was established, he stated: "Among the many and varied questions that anthropology embraces in its vast field, there is probably none of greater interest to us than that of the origins of our nation" (translation, Broca, 1860: 1). Society's, but foremost Broca's, focus had shifted. It was no longer about categorizing "foreign," "non-European" people, but about one's own "nation," about France. He dealt extensively with pre-history, the brain development of the people of France, and the mixture of "races." Where did the "nation" originate, and what "races" did it consist of? To solve these problems, he primarily relied on Blumenbach, Morton, and the Swedish anatomist Anders Retzius.

In 1862, Broca rightly stated to his fellows in the Anthropological Society: "I will limit myself to saying just one thing, and that is that if Germany has had Blumenbach, and if England has had Prichard, America has had Morton, and so far French anthropology has no name that it can put on a par with those" (translation, Broca, 1862b: 423).[15] This section shows that he would become this person. Although he moved far away from Blumenbach's monogenetic "race" theory, he nevertheless relied on several aspects of it and modified others, always praising his work as a starting point. Above all, he adopted Blumenbach's working practices: "collecting" and comparing.

Until Broca's osteological efforts, there was no comparable collection of skulls in Paris. Georges Cuvier had already examined some heads and mummies but focused primarily on zoology and paleontology. In 1861, a coincidence brought Broca a large quantity of human bones. Construction work in the center of Paris unearthed many skulls from the twelfth century. This find raised his hopes of being able to make statements about the original character and development of the French "race." Broca wanted to analyze this in greater depth using skulls from other centuries. He obtained permission to look for suitable skulls from excavations and cemetery relocations in Paris. From the *Cimetière des Innocents*, he obtained over one hundred skulls dating back to the eighteenth century. He also "collected" and examined many skulls from the nineteenth century from the West Cemetery in Paris, which gave him three large comparable series (on the significance of series in anthropological collections see Dias, 2012). In order to generate meticulously comparable measurement data from skulls, Broca needed a large number of *comparata*. In contrast to Blumenbach and Morton, he traveled to cemeteries and excavation sites and dug up skulls with his own hands or was involved in state-authorized excavations by Napoleon III.

Not only did he receive skulls from Paris, but in 1862, he traveled with his Madrid colleague, Pedro Gonzáles Velasco, to the Basque Country in Zarautz to excavate skulls. Due to lacking authorization, they worked at night and in haste. Gonzáles had first sent Broca a Basque skull that showed dolichocephalic features, which surprised Broca. The Swedish anatomist Anders Retzius had attributed a brachycephalic form to the autochthonous Stone Age "races" in Europe, like the Basques.[16] Broca wanted to get to the bottom of this problem and "collected" further skulls from the Basque

[15] Prichard later began amassing skulls, but his focus was on linguistics.

[16] Retzius was especially interested in the difference between the Caucasian Swedes and the Sámi, focusing on European "races." For Retzius' theory and his different practices of comparing see

countries in Spain and France. This shows that research questions drove his "collecting" practices, and that the representative curiosity element in Blumenbach's collection had disappeared.

Blumenbach's and Morton's "collecting" practices shifted from an *armchair community* to field research as Broca dug up skulls himself. This should not obscure the fact that the members of the Anthropological Society itself continued with established "collecting" practices. In overseas and colonized territories they continued to procure skulls, from the Canary Islands or Vietnam, through their scientific network, or they borrowed several skulls from England, e.g., through Thomas Henry Huxley. Amassing a collection was now in the hands of many persons (on the rise of field research in French anthropology around 1900 see Conklin, 2013).

Not just "collecting" changed, but with a different context and epistemic focus also comparing. Broca had a strong interest in mathematics since childhood, which he now applied to his skull studies. He studied all the craniometry publications from abroad and encountered various problems. He concluded that it was not craniometry itself that was the problem but the methodology. Craniometry was needed to understand the manifold differences between skull types and brain development. He thus explicitly turned away from Blumenbach's visual methodology. Although he praised the *norma verticalis* as the first systematic attempt, he criticized it for being based on cranioscopic—i.e., optical—methods. As it was based on measurements, Broca saw use in Camper's *facial angle* but modified the marking points. He accused Blumenbach and Camper of arbitrary selection and primarily criticized them for establishing pattern skulls, supposedly the most typical. Although this fulfilled a picturesque purpose, it was scientifically incorrect: "Beauty can't be measured or defined; it can be felt, and everyone feels it in their own way. What is the most beautiful human type? Everyone is inclined to think it's his or her own race" (translation, Broca, 1867: 630). A typical skull could only be produced by statistical averaging, excluding aesthetics from "racial" classifications. Since aesthetics were deeply intertwined with religious assumptions, Broca concluded that both had no scientifically valid function and disappeared from his scientific explanations. For Broca, the problem of craniology lay in the inductive determination of characteristics.

Whereas Blumenbach's classification included ideal types that represent a specific geographical location, Broca's categories only gained their validity through a multitude of "objects" and averages. Retzius, primarily interested in the European "races," already determined averages using the cephalic index but had too few objects; it was not representative. The *indice céphalique* or skull index, introduced by Retzius, determined the ratio of the maximum width and the maximum length of the skull, so the skulls could then be categorized as either long-skulls or short-skulls. The skull form of the autochthonous "races" in Europe troubled Broca. He criticized Retzius for establishing too simple a classification; therefore, he extended this classification of skull forms into five by dividing each type (brachycephaly (short/round) and dolichocephaly (long)) into two subtypes, adding a middle form and

---

Wilckens (2022b). In a letter from 1845, Retzius had already introduced his skull forms to Morton. They both exchanged skulls and racial theories.

introducing boundary values.[17] This division revealed a relative arbitrariness of the limit values and an overemphasis on the most minor numerical difference. Even if Broca rejected Blumenbach's optic and idealized system, he was convinced of its symmetric ordering, as Broca's division into five was not logically derived from data. Like Blumenbach, he introduced a fivefold scheme for ordering skull forms (Broca, 1861a). Although he rejected Blumenbach's aesthetic and religious arguments as explanans, he could not wholly detach himself from symmetry. The skull index must also be understood as a measurable extension of Blumenbach‘s *norma verticalis* to determine the skull's form. Broca concluded that the autochthones "races" of Europe had different skull forms, with a striking dominance, in opposition to Retzius' theory, of the dolichocephalic.

Broca's other main interest in skulls concerned the brain's capacity, closely linked to the form of the skull and its development over centuries. The obsession with skulls during this time would not be justified, according to Broca, "if the examination of the bones of the head had no more than a purely anatomical significance, and if we did not hope to find in them some data relating to the intellectual value of the various human races" (translation, Broca, 1861b: 139). The skull was only usable as a research object if intellect could be determined, following Morton's path (but rejecting phrenology).

After trying diverse skull capacity measurement methods for his Parisian skull series, he decided to follow Morton's method of filling the skull with lead shot using a cylinder, but improved it by inserting a fine rod to spread the shot throughout the skull. He established the *méthodes moyennes* by drawing the average from comparable skull variables (Broca, 1879; Blanckaert, 1991). Thereby, a whole series of skulls built a *comparatum*, whereas Blumenbach mostly used singular skulls as *comparatum*. To this end, Broca developed a number of new measurement apparatuses, including a craniograph or a better facial goniometer, and compiled a wealth of statistical tables. Similar to Morton, the skull's volume was of great importance, but for Broca, the size of the frontal cerebral region crystallized as the hallmark of high intelligence when his findings did not show his expected results. When the general average of cranial capacity contradicted his presuppositions—namely that the skull's capacity had increased over the centuries and that the French, especially the Parisians, had the largest cranial volume—he chose the frontal cerebral region as central *tertium*, as it was most pronounced among Parisians and included other *tertia*, such as the richness of the brain windings, the shape, and the relative degree of development of the different brain regions. Another example of a *tertium* change lay in the inclusion of gender variables. Unlike Morton and Blumenbach, Broca included sexual differences in his research and used them to explain unexpectedly smaller brains in groups of people. Even though Blumenbach had two female pattern skulls, he did not include

[17] These types include *dolichocéphales purs, sous-dolichocéphales, brachycéphales purs, sous-brachycéphales, mésaticéphale.*

gender differences in his interpretations. Female types in Blumenbach's work functioned only abstractly and in connection with aesthetically ascribed characteristics, but not as concrete individual personalities.[18]

Broca compared the Parisian series with finds from the Paleolithic and Neolithic periods to investigate brain development and explain the composition and origin of the French. The finds of the so-called *Homme-Mort* from the Neolithic period and the even older *Cro-Magnon* from the Quaternary period served as *comparata*, with a dolichocephalic skull shape. It is not surprising that he attributed the dolichocephalic form to prehistoric humans, as, for him, the frontal cerebral region was a marker of intelligence, foremost indicated through the brachycephalic form. The prehistoric people on French territory would have already had roughly the same skull shape as the later Basque-Spanish and African populations. Broca thus perpetuated widespread stereotypes of "backwardness" and "primitiveness." The *Homme-Mort* people had been extinguished in the Neolithic period, which was perfectly compatible with his species concept. Prehistory had a hinge function in his research and could link many of his phenomena (on the institutional rise of prehistory see Richard, 1995; Kaeser, 2001).

He always emphasized an unbiased approach to the data to achieve the most significant possible degree of objectivity; in the spirit of science, he always considered falsifiability, primarily through new *comparata*. However, what was striking and what contradicted his claims of scientific rigor were the many *tertia* changes. He looked for new *tertia* if the data did not fit the assumed results (Gould, 1981). Unlike Morton, he did not change the data or samples but looked for further, different data or criteria. Examples included the change from the overall cranial volume to the anterior part of it or the inclusion or exclusion of female data in his explanations.

Although Broca no longer followed continental ascriptions, he could not break entirely away from Blumenbach's and other's labeling. Broca was sharply critical of the so-called "pentagénistes" who favored polygenism but adopted Blumenbach's continental fivefold scheme, as did Morton. However, Morton later became more doubtful about this assumption and advocated more original "races." Broca assumed that there were originally many different human species at different times. However, they were hardly traceable due to hybridization, i.e. the mixing of "races." He developed his hybridity theory by the end of the 1850s, extending Morton's theory on hybridity. The starting point was a Léporid, a fertile breed of hare and rabbit, which Alfred Roux bred in Angoulême. Broca rejected Buffon's species concept and questioned, in general, the comprehensibility of species.[19] Consequently, he rejected typologies and general classifications as abstractions and simplifications. Because

[18] This becomes apparent in his *Abbildungen naturhistorischer Gegenstände*, in which he chose five men to depict his main varieties.

[19] Fertile offspring would only mean the compatibility of the genitals. He differentiated between unilateral and bilateral hybridity. He adopted Morton's four degrees of hybridity and categorized them into two classes: *inférieur* and *supérieur*. There was also the possibility of *croisement de retour*. In contrast to Morton, he was convinced that all "races" were hybrid "races." The French nation had, according to Broca, emerged from a superior cross-breeding of Celts and Kimris.

he rejected ideal types, Broca rarely spoke of continental groups of people. Nevertheless, he assumed persistent archetypes, even though he was convinced of the emergence of new "races" through interbreeding and some species' possible extinction. One should not confuse type with the idea of "race," for which a character that is inherited and sufficiently stable is necessary. "Races" would divide into a smaller, compartmentalized number of local groups, and the ensemble of their common characteristics would form an anthropological type. Broca saw a similarity in the statics and dynamics of the brains of people of the same type. For him, there were only three of these anthropological types, which he adopted from Blumenbach: Ethiopian, Mongolian, and Caucasian, but he hardly used this superordinate categorization in his research.

Broca's research was deeply embedded in French nation-building after turbulent and revolutionary times—from the French Revolution to the July monarchy, the Second Republic, the Second Empire, and the Third Republic. Furthermore, the state needed a scientifically valid narrative about becoming a nation, as state-founded excavations showed. During this time, the narrative of the French nation, which was believed to originat from the Celts, was state-driven. Broca's research showed clearly that nationalism (in connection with the emergence of nation-states/modern nationalism) had a decisive impact on anthropological research (Blanckaert, 2001). Whereas previously, "the others" had been the focus of science, it was now about one's identity and nation. Broca was an ardent Republican who later also became politically active. Although he strongly supported the American School's thesis and a hierarchy of "races," he distanced himself from slavery, which had already been abolished in France and was not in keeping with republican ideals.

4.4 Conclusion

Different understandings of science and contexts have led to changed theories of "race"—simultaneously to their stability and intractability. Not only Blumenbach, although central, but also a transnational community of practice, has led to the perpetuation of "race" theories through contextual adjustments. Comparing and "race" theories can be easily dis- and re-embedded in different contexts. This analysis has shown how various contexts and the presence of others, e.g., the Enlightenment, colonialism, the removal of Indigenous people, slavery, and nation-building, led to nuanced changes in the methodology and understanding of "race" because of their specific purposes.

Blumenbach's skull-based racial theory became the benchmark for all subsequent comparative anatomists. It is remarkable that despite a completely different, polygenic theory of "race," his fivefold classification, his labels, and his empirical foundation have endured. Even when deviations from the fivefold scheme and his concept of "race" have occurred, the symmetrical approach, the reliance on skulls, and the constant reference to Blumenbach have remained. Morton and Broca adopted his practices of "collecting" and comparing skulls but added measurement, due to new scientific ideals and understandings. Among all three anatomists, their "collecting" and comparing influenced each other strongly. The contingency of theories became apparent, especially when looking at their interplay. Measuring stabilized practices of comparing skulls and enabled it to be strengthened for at least the next hundred years. By distinguishing between different practices, an analytical gain becomes visible, especially the change in racial construction processes at different levels.

A specific look at practices of comparing can show the ties to Blumenbach's approach and the respective racial constructions. Comparison also reveals much about the context of larger dynamics and epistemologies. Blumenbach made the skull the main comparative criterion. In the comparative arrangement, it became the *comparatum*. The *tertia*, on the other hand, changed constantly since then: from aesthetics and symmetry to cranial and anterior capacity. Blumenbach's descriptive comparisons became numeric and quantitative. Thus, the use of skull comparisons had been constant since Blumenbach, but the practices of comparing had changed. As they changed due to different contexts and epistemologies, conceptions of "race" also changed, guaranteeing their persistence. While Blumenbach primarily made a continental classification, "race" later increasingly shifted toward national and regional categories.

Blumenbach had anti-racist intentions by pursuing idealistic Enlightenment approaches and pleaded for the intellectual unity of humankind, but still propagated racialized stereotypes and an aesthetic hierarchy. His approach performed well in racist and hierarchical contexts. The unintended consequences in the reception of his classification have been considerable. They have undoubtedly had an enormous impact on "racial" imaginings to this day. It was only through them that Blumenbach's classification was able to achieve such popularity in the first place.

References

Armstrong-Fumero, F. (2014). "Even the most careless observer": Race and visual discernment in physical anthropology from Samuel Morton to Kennewick Man. *American Studies, 53*(2), 5–29.

Bitter-Smirnov, S. (2022). Arguing against monogenism: Strategies in polygenist argumentation against the unity of humankind. In M. S. Wilckens & J. T. D. Gärtner (Eds.), *Racializing humankind: Interdisciplinary perspectives on practices of 'race' and racism* (pp. 233–255). Böhlau

Blanckaert, C. (1991). Méthode des moyennes et notion de "Série Suffisante" en Anthropologie Physique: 1830–1880. In J. Feldman, G. Lagneau & B. Matalon (Eds.), Moyenne, milieu, centre. Histoires et usages. *Histoire des Sciences et des Techniques* (Vol. 5, pp. 213–243). École des Hautes Études en Sciences Sociales.

Blanckaert, C. (2001). Les usages de l'anthropologie. In ibid. (Ed.). Les politiques de l'anthropologie. Discours et pratiques en France (1860–1940). *Collection Histoire des sciences humaines* (pp. 9–26). L'Harmattan.

Blanckaert, C. (2009). De la race à l'évolution. Paul Broca et l'anthropologie française (1850–1900). *Histoire des sciences humaines*. L'Harmattan.

Blumenbach, J. F. (1789). Über Menschen-Racen und Schweine-Racen. *Magazin für das Neueste aus der Physik und Naturgeschichte, 6*, 1–13.

Blumenbach, J. F. (1795). *De generis humani varietate nativa. Editio tertia.* Vandenhoeck & Ruprecht.
Blumenbach, J. F. (1798). *Über die natürlichen Verschiedenheiten im Menschengeschlechte.* Breitkopf & Härtel.
Boas, F. (1896). The limitations of the comparative method of anthropology. *Science, 103*(4), 901–908.
Bodenmann, S., & Rey, A. -L. (Eds.). (2018). What does it mean to be an empiricist? Empiricism in Eighteenth-Century sciences. *Boston studies in the philosophy and history of science 331.* Springer.
Böker, W. (2018). Blumenbach's collection of human skulls. In N. Rupke, & G. Lauer (Eds.). *Johann Friedrich Blumenbach: Race and natural history, 1750–1850* (pp. 80–95). Routledge.
Boulter, S. (2021). Evolution and the principle of proportionality. In W. M. R. Simpson, R. C. Koons & J. Orr (Eds.), *Neo-Aristotelian metaphysics and the theology of nature* (pp. 125–148). Routledge.
Bourdieu, P. (2014). *Sozialer Sinn: Kritik der theoretischen Vernunft.* Suhrkamp.
Broca, P. (1860). Recherches sur l'ethnologie de la France. In Société d'Anthropologie de Paris (Ed.), *Mémoires de la Société d'Anthropologie de Paris* (1) 1860–63 (pp. 1–54).
Broca, P. (1861a). Sur des crânes provenant d'un cimetière de la cité. Antérieur au troisième siècle. In Société d'Anthropologie de Paris (Ed.), *Bulletins de la Société d'Anthropologie* (pp. 501–513).
Broca, P. (1861b). Sur les volumes et la forme du cerveau suivant les individus et suivant les races. In Société d'Anthropologie de Paris (Ed.), *Bulletins de la Société d'Anthropologie* (pp. 139–207).
Broca, P. (1862a). La linguistique et l'anthropologie. In Société d'Anthropologie de Paris (Ed.). *Bulletins de la Société d'Anthropologie* (pp. 264–319).
Broca, P. (1862b). Discussion sur le rapport de M. Dally concernant les Américains. In Société d'Anthropologie de Paris (Ed.), *Bulletins de la Société d'Anthropologie* (pp. 417–437).
Broca, P. (1867). Discussion. In Société d'Anthropologie de Paris (Ed.), *Bulletins de la Société d'Anthropologie de Paris* (pp. 629–631).
Broca, P. (1879). Méthodes des moyennes. Étude des variations craniométriques et de leur influence sur les moyennes; détermination de la série suffisante. In Société d'Anthropologie de Paris (Ed.), *Bulletins de la Société d'Anthropologie de Paris* (pp. 756–820).
Candea, M. (2018). *Comparison in anthropology: The impossible method.* Cambridge University Press.
Colavito, J. (2020). *The mound builder myth: Fake history and the hunt for a "lost white race".* University of Oklahoma Press.
Collins Cook, D. (2006). The old physical anthropology and the new world: A look at the accomplishments of an antiquated paradigm. In J. E. Buikstra & L. A. Beck (Eds.), *Bioarcheology* (pp. 27–71). Elsevier.
Conklin, A. L. (2013). *In the museum of man: Race, anthropology, and empire in France, 1850–1950.* Cornell University Press.
Conrad, S. (2012). Enlightenment in global history: A historiographical critique. *American Historical Review, 117*, 999–1027.
Daston, L., & Galison, P. (2007). *Objectivity.* Princeton University Press.
De Angelis, S. (2021). Frühneuzeitliche Anthropologie: Grundaspekte, Konzepte, Entwicklungslinien zwischen Renaissance und Spätaufklärung. Ein Überblick. In S. Salatowsky, W. Schmidt-Biggemann, & J. -L. Albrecht (Eds.), *De homine: Anthropologie in der Frühen Neuzeit* (pp. 15–29). Franz Steiner Verlag.
De Boer, K., & Prunea-Bretonnet, T. (Eds.). (2021). The experiential turn in Eighteenth-Century German philosophy. *Routledge studies in Eighteenth-century philosophy.* Routledge.
Dias, N. (2012). Nineteenth-century French collections of skulls of the cult of bones. *Nuncius, 27*, 330–347.
Dougherty, F. W. P. (1996). *Gesammelte Aufsätze zu Themen der klassischen Periode der Naturgeschichte.* Norbert Klatt Verlag.

Doughtery, F. W. P. (1985). Johann Friedrich Blumenbach und Samuel Thomas Soemmerring: Eine Auseinandersetzung in anthropologischer Hinsicht? In G. Mann, & F. Dumont (Eds.), *Samuel Thomas Soemmerring und die Gelehrten der Goethezeit*. Schwabe.

Epple, A. (2015). Doing Comparisons—Ein praxeologischer Zugang zur Geschichte der Globalisierung/en. In ibid., & W. Erhart (Eds.). *Die Welt beobachten. Praktiken des Vergleichens* (pp. 161–199). Campus.

Epple, A. (2020). Inventing white beauty and fighting black slavery: How Blumenbach, Humboldt, and Arango y Parreño contributed to Cuban race comparisons in the long nineteenth century. In ibid., W. Erhard, & J. Grave (Eds.). *Practices of comparing: Towards a new understanding of a fundamental human practice* (pp. 295–328). Bielefeld University Press.

Epple, A., Erhart, W., & Grave, J. (Eds.). (2020). *Practices of comparing*. Bielefeld University Press.

Fabian, A. (2010). *The skull collectors: Race, science, and America's unburied dead*. The University of Chicago Press.

Fredrickson, G. M. (1971). *The black image in the white mind: The debate on Afro-American character and destiny, 1817–1914*. Harper & Row.

Gerbner, K. (2019). *Christian slavery: Conversion and race in the protestant Atlantic world*. University of Pennsylvania Press.

Giddens, A. (1984). *The constitution of society*. University of California Press.

Giddens, A. (1990). *The consequences of modernity*. Stanford University Press.

Golf-French, M. (2023). Teaching race in German Enlightenment: Christoph Meiners' history of humanity in institutional context. In A. Beeton, E. P. Bernstein, E. Kent, & R. Winkler (Eds.),*The mind is its own place? Early modern intellectual history in an institutional context*. Oxford University Press.

Gould, S. J. (1978). Morton's ranking of races by cranial capacity. *Science, 200*, 503–509.

Gould, S. J. (1981). *The mismeasure of man*. Norton.

Hüning, D. (2023). Christoph Meiners und die Ursprünge des Rassismus im Zeitalter der Aufklärung. In S. Klingner & G. Stiening (Eds.), *Christoph Meiners (1747–1810)* (pp. 73–94). De Gruyter.

Hünniger, D. (2018). Bilder machen—Charaktere, Stereotype und die Konstruktion menschlicher Varietät bei Johann Friedrich Blumenbach. In ibid, & D. L. Eudell (Eds.), *Lichtenbergs MenschenBilder. Charaktere und Stereotype in der Göttinger Aufklärung* (pp. 65–77). Göttinger Verlag der Kunst.

Junker, T. (2018). Blumenbach's theory of human races and the natural unity of humankind. In N. Rupke, & G. Lauer (Eds.), *Johann Friedrich Blumenbach: Race and natural history, 1750–1850* (pp. 96–112). Routledge.

Kaeser, M. -A. (2001). L'Internationalisation de la préhistoire, une manoeuvre tactique? Les conséquences épistémologiques de la fondation des congrès internationaux d'anthropologie et d'archéologie préhistoriques. In C. Blanckaert (Ed.), *Les politiques de l'anthropologie: Discours et pratiques en France (1860–1940)* (pp. 201–230).

Kassabaum, M. C. (2021). *A history of platform mound ceremonialism. Finding meaning in elevated ground*. University Press of Florida.

Keel, T. (2013). Religion, polygenism and the early science of human origins. *History of the Human Sciences, 26*(2), 3–32.

Keel, T. (2018a). *Divine variations*. Stanford University Press.

Keel, T. (2018b). Blumenbach's race science in the light of Christian supersessionism. In N. Rupke, & G. Lauer (Eds.), *Johann Friedrich Blumenbach: Race and natural history, 1750–1850* (pp. 123–141). Routledge.

Klatt, N. (2008). *Kleine Beiträge zur Blumenbach-Forschung 1*. Norbert Klatt Verlag.

Klatt, N. (2010). *Kleine Beiträge zur Blumenbach-Forschung 3*. Norbert Klatt Verlag.

Lauer, G. (2014). Über Menschen, Schweine und andere Rassen: Die Varietäten der Menschen und ihre zoologische Kritik in der gelehrten Literatur des 18. Jahrhunderts. In C. Gantet, & F. Schock

(Eds.). Zeitschriften, Journalismus und gelehrte Kommunikation im 18. Jahrhundert. Festschrift für Thomas Habel. *Presse und Geschichte—Neue Beiträge 81* (pp. 221–232). Edition Lumière.
Lewis, J. E., DeGusta, D., Meyer, M. R., Monge, J. M., Mann, A. E., Holloway, R. L. (2011). The mismeasure of science: Stephen Jay Gould versus Samuel George Morton on skulls and bias. *PLoS Biology, 9*(6).
Malcolmson, C. (2013). *Studies of skin color in the early Royal Society: Boyle, Cavendish, Swift*. Taylor & Francis Ltd.
Mann, B. A. (2003). *Native Americans, archeologists, & the mounds*. Peter Lang Publishing.
Mazzolini, R. G. (2014). Skin color and the origin of physical anthropology (1640–1850). In S. Lettow (Ed.). *Reproduction, race, and gender in philosophy and the early life sciences* (pp. 131–161). State University of New York Press.
McMahon, R. (2018). The history of transdisciplinary race classification: Methods, politics and institutions, 1840s–1940s. *The British Journal for the History of Science , 51*(1), 41–67.
Meigs, C. D. (1851). *A memoir of Samuel George Morton, M. D., late president of the Academy of Natural Sciences of Philadelphia*. T.K. & P.G. Collins.
Meiners, C. (1785). *Grundriß der Geschichte der Menschheit*. Verlag der Meyerschen Buchhandlung.
Michael, J. S. (1988). A new look at Morton's craniological research. *Current Anthropology, 29*(2), 349–354.
Michael, J. S. (2017). Nuance lost in translation. *N.T.M. 25*, 281–309.
Michael, J. S. (2020). An "American Humboldt"? Memorializing philadelphian physician and race supremacist Samuel George Morton. *Pennsylvania History: A Journal of mid-Atlantic Studies, 87*(2), 279–312.
Michael, J. S. (2021). Porträts von interessanten Personen: A new look at J. F. Blumenbach's typological labels and the exemplars he discussed in his anthropological research. In Deutsche Gesellschaft für Geschichte und Theorie der Biologie (Ed.), *Annals of the history and biology* (Vol. 25, pp. 65–101). Universitätsverlag Göttingen.
Morton, S. G. (1839). *Crania Americana: Or, a comparative view of the skulls of various aboriginal nations of North and South America: To which is prefixed an essay on the varieties of the human species*. J. Dobson & Simpkin, Marshall & co.
Morton, S. G. (1844a). *An inquiry into the distinctive characteristics of the aboriginal race of America*. John Penington.
Morton, S. G. (1844b). *Crania Aegyptiaca: Or, observations of Egyptian ethnography, derived from anatomy, history and the monuments*. John Penington.
Morton, S. G. (1847). *Hybridity in animals and plants, considered in reference to the question of the unity of human species*. B. L. Hamlen.
Painter, N. I. (2010). *The history of white people*. W.W. Norton & Company.
Parfitt, T. (2020). *Hybrid hate: Conflations of antisemitism & anti-Black racism from the Renaissance to the Third Reich*. Oxford University Press.
Patterson, H. S. (1854). *Memoir of the life and scientific labors of Samuel George Morton, M. D.* Lippincott, Grambo.
Poskett, J. (2015). National types: The transatlantic publication and reception of Crania Americana (1839). *History of Science, 53*(3), 264–295.
Radhakrishnan, R. (2009). Why compare? *New Literary History, 40*(3), 453–471.
Reckwitz, A. (2002). Toward a theory of social practices: A development in culturalist theorizing. *European Journal of Social Theory, 5*(2), 243–263.
Richard, N. (1995). La naissance de la préhistoire. In C. Blanckaert, J.-L. Fischer & R. Rey (Eds.), *Nature, histoire, société. Essais en hommage à Jacques Roger* (pp. 419–431).
Rupke, N., & Lauer, G. (Eds.). (2018). *Johann Friedrich Blumenbach: Race and natura history, 1750–1850*. Routledge.
Sloan, P. R. (2002). Preforming categories: Eighteenth-century generation theory and the biological roots of Kant's A Priori. *Journal of the History of Philosophy, 40*(2), 229–253.

Sommer, M. (2024). *The diagrammatics of 'race': Visualizing human relatedness in the history of physical, evolutionary, and genetic anthropology, ca. 1770–2020*. Open Book Publishers.

Stanton, W. (1960). *The leopard's spots: Scientific attitudes towards race in America, 1815–1859*. University of Chicago Press.

Weisberg, M. (2014). Remeasuring man. *Evolution & Development, 16*(3), 166–178.

Wilckens, M. S. (2022a). Differenz messen. Der Schädel als 'Objekt-Subjekt' zur Klassifikation von "Rassen" in der Vergleichenden Anatomie des 18. und 19. Jahrhunderts. In C. Aust, A. Flüchter, & C. Jarzebowski (Eds.). *Verglichene Körper. Normieren, Urteilen, Entrechten in der Vormoderne* (pp. 179–203). Franz Steiner Verlag.

Wilckens, M. S. (2022b). Racializing, collecting and comparing skulls: Johann Friedrich Blumenbach's and Anders Retzius' practices of ordering humankind. In ibid., & J. T. D. Gärtner (Eds.), *Racializing humankind: Interdisciplinary perspectives on practices of 'race' and racism* (pp. 91–110). Böhlau.

Wilckens, M. S., & Gärtner, J. T. D. (2022). Introduction—conceptualizations and interdisciplinary perspectives on practices of 'race' & racism. In ibid. (Eds.), *Racializing humankind: Interdisciplinary perspectives on practices of 'race' and racism* (pp. 9–30). Böhlau.

Wilckens, M. S., & Kurzwelly, J. (2022). Wert und Verwendung menschlicher Überreste: Vergangene und gegenwärtige Perspektiven im interdisziplinären Dialog. *Historische Anthropologie, 30*(3), 329–349.

Winterer, C. (2012). Where is America in the republic of letters? *Modern Intellectual History, 9*(3), 597–623.

Winterer, C. (2016). *American Enlightenments: Pursuing happiness in the age of reason*. Yale University Press.

Winterer, C. (2024). *How the New World became old: The deep time revolution in America*. Princeton University Press.

Wolff Mitchell, P., & Michael, J. S. (2019). Bias, brains, and skulls: Tracing the legacy of scientific racism in the nineteenth-century works of Samuel George Morton and Friedrich Tiedemann. In J. A. Thomas, & C. Jackson (Eds.), *Embodied difference: Divergent bodies in public discourse* (pp. 77–98). Rowman & Littlefield.

Zammito, J. H. (2018). The rise of paleontology and the historicization of nature: Blumenbach and Deluc. In N. Rupke & G. Lauer (Eds.), *Johann Friedrich Blumenbach: Race and Natural History, 1750–1850* (pp. 197–232). Routledge.

Primary Sources

Banks, J. to Blumenbach. May 18, 1790. No. 613, p. 306f. F. W. P. Dougherty (Ed). The Correspondence of Johann Friedrich Blumenbach, Vol. III (2010). Norbert Klatt.

Blumenbach, J. F. to Banks. January 8, 1794. No. 826, p. 295–298. F. W. P. Dougherty (Ed). The Correspondence of Johann Friedrich Blumenbach, Vol. IV (2012). Norbert Klatt.

Blumenbach, J. F. to Banks. March 10, 1794. No. 838, p. 312f. Ibid.

Blumenbach, J. F. to van Geuns. July 16, 1795. No. 927, p. 437f. Ibid.

Blumenbach, J. F. *Über die einzelnen Schädel*, 61 Faszikel. Cod. Ms. Blumenbach V, Niedersächsische Staats- und Universitätsbibliothek Göttingen (SUB).

Brodersen, J. to Loretz. September 17, 1792. No. 772, p. 197–199. F. W. P. Dougherty (Ed). The Correspondence of Johann Friedrich Blumenbach, Vol. IV (2012). Norbert Klatt.

Gliddon, G. to Morton. March 31, 1839. Samuel George Morton Papers (SGMP), American Philosophical Society (APS).

Morton, S. G. Notes. Ca. 1842. Crania Aegyptiaca, research and notes. Vol. 3. SGMP, Library Company Philadelphia (LCP).

Morton, S. G. to Ruschenberger. February 28, 1832. Morton Letter Book, 1832–1837, C0199 no. 737q, Special Collection, *Princeton University Library*.

Morton, S.G. to Gliddon. November 2, 1837. Cit. after Patterson. Memoir of the Life and Scientific Labors of Samuel George Morton, S. xxxvf.
Richard, R. S. Secretary. Newspaper clippings re. Dr. Morton's death. SGMP, APS.
Townsend, J. K. to Morton. September 20, 1835. SGMP, APS.
Von Asch, G. T. to Blumenbach. May 29, 1793. No. 811, p. 256f. F. W. P. Dougherty (Ed). The Correspondence of Johann Friedrich Blumenbach, Vol. IV (2012). Norbert Klatt.
Wood, W. to Drake. September 28, 1833. SGMP, APS.
Zu Wied-Neuwied, M. an Morton. May 2, 1837. SGMP, APS.

Malin S. Wilckens is a postdoctoral researcher at the Leibniz Institute of European History in Mainz and has been a joint visiting postdoctoral research fellow at the German Historical Institute London and the Institute of Advanced Studies at University College London. She previously worked as a research assistant for 19th and 20th-century history at the universities of Bielefeld and Kiel. She was a doctoral fellow of the German Academic Scholarship Foundation at Bielefeld University, with research stays at Stanford University and the German Historical Institute in Washington. Her research interests include the history of science and knowledge, historical 'race' studies, global history, and the history of the environment and technology.

Chapter 5
Blumenbach's Malay Variety: Language and Emotionality in a Transitory "Race"

Ivana Pražić, Shakila Che Dahalan, Khairunnisa A Shukor, and Sara Shakilla Mohd Salim

Abstract This chapter explores the conceptual genesis of Johann Friedrich Blumenbach's Malay "race." Introduced at the end of the eighteenth century, Blumenbach's Malay variety stood out from his other human varieties by being named after a language rather than a place. To understand this taxonomic exceptionality, we trace the forms of argumentation he deployed to gradually include a fifth variety to an initial four-fold division and to name it after the Malay language. This investigation yields insight into Blumenbach's conceptual, narrative, and methodological dependence on (then) contemporary and earlier studies of human origins and kinship ties through studies of language. Finally, we analyze the epistemic and cultural legacy of his five-fold division of human species and his ranking of the Malay variety as being in transition between the "primordial" European and the "extreme" African variety. In order to identify the historical effects of Blumenbach's race science on present-day, insular Southeast Asia, we rely on critical studies of "race" informed though affect theory in particular. We deploy Sara Ahmed's concept of emotionality to understand how racialized qualifiers of transitoriness had been reconfigured as a colonial pedagogic strategy enabling the approximation of the Native American to the Filipino and the Malay.

Keywords Malay language · Malay race · Malay human variety · Race (as language · biological · cultural) · Transitoriness/transience of race · Emotionality · Linguistics · Colonial education

I. Pražić (✉) · S. C. Dahalan · K. A. Shukor · S. S. M. Salim
The Faculty of Human Sciences, Sultan Idris Educational University (UPSI), Tanjung Malim, Malaysia
e-mail: iprazic@gmail.com

V. Shmidt and S. De Angelis (eds.), *Blumenbach's Racial Classification*, History, Philosophy and Theory of the Life Sciences 39,
https://doi.org/10.1007/978-3-032-15950-2_5

5.1 Introduction

Malaysia and Brunei Darussalam are two Southeast Asian nation-states with a Malay majority population. Populations of Malay descent also live in Indonesia, Singapore, Thailand, the Philippines and Vietnam, in varying numbers. The multicultural composition of Malaysia today, the standard story goes, is a result of British colonialism, which relied on immigrant laborers from India and China to cultivate cash crops, extract tons of rubber, mine tin ore, and otherwise facilitate exploitation of the lands where Malay was the majority tongue. Despite the post-Second World War Western-led tendency to avoid the term "race,"[1] in Malaysia, Singapore, and Brunei Darussalam, it is still in use as a non-derogatory term. A government portal explains that Malaysia is "[a] multiracial and multicultural nation in which Malays, Indian, Chinese and people of various other ethnicities live together in peace and harmony." (Government of Malaysia's Official Gateway, 2025). The same text in Malay does not use the term *ras* (the Malay-language translation of "race"). Instead, it uses the terms *kaum* and *bangsa*. The former stands for "race/culture," while the latter stands for "nation," whereas ethnicity is translated as (*kumpulan*) *etnik*, or ethnic group.[2]

In the history of European scientific classification, Malay as one of the five varieties of humankind is credited to Johann Friedrich Blumenbach (1795; henceforth: *De Generis 3*). Nonetheless, the Malaysian educational system does not mention Blumenbach and his celebrated human taxonomy. This didactic lack represents one of our motives for undertaking research on Blumenbach's racial taxonomy. Does Blumenbach's definition of the Malay people in biological and physiological, but also wider inter-relational, terms have any bearing on how Malays are (self)imagined today? How did Blumenbach acquire his knowledge of the Malay people? Were there any traces of Blumenbach's race science in the colonial racist politics in Malaya and elsewhere? This chapter seeks to answer these questions.

---

[1] On the topic of the anthropological role of turning culture into a carrier of racial difference after the Second World War, see for instance Kamala Visweswaran, *Un/common Cultures: Racism and the Rearticulation of Cultural Difference* (Duke University Press, 2010).

[2] For an inspiring discussion on the versatility of Malay-language terms for "race," and the overlapping meanings of *bangsa* with "nation" *and* "race," see Sandra Khor Manickam, "Common ground: Race and the Colonial Universe in British Malaya," *Journal of Southeast Asian Studies* 40, no. 3 (October, 2009): 594–5. https://doi.org/10.1017/S0022463409990087. Death certificates in Malaysia indicate *keturunan* (offspring, ancestry, heredity) as the equivalent for "race/ethnicity." According to Judith Nagata, the term *bangsa* originally stood for "community," but in colonial times, it was recoded to (imprecisely) denote European concepts of "race," "ethnicity," and "nation." Judith Nagata, "Boundaries of Malayness: 'We Have Made Malaysia: Now It Is Time to (Re)Make the Malays but Who Interprets the History'?" in *Melayu: Politics, Poetics and Paradoxes of Malayness*, ed. Maznah Mohamad and Syed Muhd Khairudin Aljunied (NUS Press, 2011), 18.

In the first section of this chapter, we explore the genesis of the concept and space for human variation, the Malay included, by analyzing Blumenbach's writings on the topic. Blumenbach only exceptionally deployed the term "race," rather than its Latin or German counterparts, to speak about human differences. Bronwen Douglas has argued that "race" was only emerging as a purely scientific/biological concept in Blumenbach's work, but it never managed to eradicate its pre-scientific genealogical connotations. In the last section of *De Generis 3*, Blumenbach replaces the hitherto ubiquitous Latin term *gens* (denoting "nation," "race," or "people") "by two botanical metaphors, *stirps* and *stemma*, both connoting descent from an ancestral stock" (Douglas, 2008, 38). Based on our research, apart from, or even more than, botany, Blumenbach was indebted to a linguistically informed genealogy of human families and variations in building his scientific study of "race." Blumenbach's exceptional mentioning of *races* (sic) in *De Generis 3* is related to his discussion of the Malay variety of one human species, which he famously introduced in the last edition of his *opus magnum* (Blumenbach, 1795, 320). In contrast to the previous four varieties named after places or geographical regions, Blumenbach names the Malay after language. Why does he include the Malay variety last? Which geographical region did Blumenbach's Malay inhabit?

As Blumenbach never traveled outside Europe, in the second section, we turn our attention to his sources of information on human variation. We identify a number of works, approaches, and scholars that left an indelible, though largely unacknowledged, epistemic imprint on Blumenbach's race science. We also discuss the role of one of the most versatile and influential intellectual figures of Blumenbach's time, the English Joseph Banks, who accompanied Captain Cook on his first voyage around the world (1768–1771) and collected specimens for subsequent research. In fact, we demonstrate that Blumenbach's Malay "race" is an excellent example of what Bill Ashcroft (2001, 311–13) has theorized as a categorical entanglement from the animal world, namely, between "race," as the biological carrier of human difference, and language, as the main site of human difference.

Our interest in the third section of the chapter turns to the cultural legacy of Blumenbachian race science in general and its Malay variety in particular, in the Southeast Asian world. Our heuristics here are organized around Blumenbach's ranking of the Malay variety as transitory, falling between the "primary" variety (European) and one of the two "extreme" varieties, the Ethiopian (referring to sub-Saharan Africa).[3] Blumenbach intimates transience as proof of racial mutability or even contingency, and over the course of the nineteenth and early

[3]According to Ricardo Roque, since its inception in the late eighteenth century until well into the twentieth century, the scientific ordering of "race" was focused on relating bodily features with supposedly unchanging inner characteristics, mental and cultural forms. Hence, most of race science or scholarship has been organized around the idea of "fixity" or the permanence of "race," although it actually depended on material circulation of bodies or parts thereof. Ricardo Roque, "Race and the Mobility of Human Things," *Science, Technology & Human Values* 39, no. 4 (2014), 608–9.

twentieth century, this feature was used as a racialized (dis)qualifier of the Malay in relation to Europeans. In order to better understand the commanding power of Blumenbach's transience in forming/dissolving barriers between collectives of human bodies grouped into "races," we approach it as an affect or emotion. In so doing, we tap into the intellectual legacy of a vast body of literature that foregrounds the importance of affect in building and maintaining colonial racial hierarchies in Southeast Asia (and elsewhere) in the nineteenth and twentieth centuries.[4] We show here that transience, as a racially formative concept drawing heuristic power from its invisible affective charge, enabled Blumenbach to disguise the contingency of and hierarchies between his varieties by turning them into a scientific rule.

We believe that it is possible to view Blumenbach as part of the vanishing "long eighteenth century" (c. 1680–1830), which Jürgen Osterhammel (2018, 22; 26–7; 32) defined as the period in which European views and attitudes on Asia reflected a certain "intellectual equilibrium" between Asia and Europe. Having said so, and despite his "anti-racist" political stance, Blumenbach's racial sensitivities did anticipate some of the colonial racial hierarchies of the nineteenth and twentieth centuries. To understand how Blumenbach ranked and grouped or divided certain populations, we comment on some of the emotionally charged descriptions of populations of people he later grouped under his Malay "race" that he had encountered and absorbed from the sources he consulted.

As Sara Ahmed has shown, "emotions are not 'in' either the individual or the social, but produce the very surfaces and boundaries that allow the individual and the social to be delineated as if they are objects" and grouped together or away from each other (Ahmed, 2014, 10). We deploy Ahmed's concept of "emotionality, which she defines "as a claim *about* a subject or a collective [...] clearly dependent on relations of power, which endow 'others' with meaning and value" (Ahmed, 2014, 4). What emotional labor maintains human variety in its transitory state?

Blumenbach deploys the trope of transitoriness, or transience, to *both* the Malay and the Native American varieties. The emotional legacy of such ranking is, as we argue here, what informed the belief in the interchangeability between Filipinos, Native Americans, and Malays in colonial educational systems for "natives" at the turn of the twentieth century. We show that colonial educational policies at the turn of the twentieth century, as informed by the United States in the Philippines and the British in Malaya, depended on a historicized emotional juxtaposition between the Native American and the Malay. Our conclusion here is that the unstudied affective legacy of Blumenbach's race science informed and

[4] For a discussion of racism and foundational colonial affective dispositions to it, see Ann Laura Stoler, *Carnal Knowledge and Imperial Power: Race and the Intimate in the Colonial Rule*, second edition (University of California Press, 2010). The massive output of critical colonial studies informed through the affective turn is impossible to cover here. Another example would be Elizabeth A. Povinelli, *The Empire of Love: Toward a Theory of Intimacy, Genealogy, and Carnality* (Duke University Press, 2006).

was disseminated through colonial educational system and special curricula for the "native" in the United States as much as in Southeast Asia. This would also explain the consequential lack of critical race theory in particular in the historical study of the region.[5]

5.2 Mapping the Malay: Southerners, Polynesians, Australasians

In the first edition of *De Generis Humani Varietate Nativa* (1776), Blumenbach describes his four human varieties, which were all numbered rather than named at this stage. The second variety spread across.

> Asia beyond the Ganges, and below the river Amoor [sic], which looks towards the south, together with the islands, and the greater part of those countries that are now called Australian. Men of dark colours, snub noses, with winking eyelids drawn outwards at the corners, scanty, and stiff hair.[6]

Hence, we can see here that Blumenbach's second variety included the peoples of mainland Central and East Asia *and* pen/insular Southeast Asia, as well as Australia. Although he mentions some Southeast Asian islanders here, there is no mention of the Malay people.[7]

In relation to the peoples he describes, the color of the skin, and the shape and size of the nose were among the distinguishing features. With regard to the "color" of people comprising this variety, Blumenbach utilized the terms *coloris fusci homines*, translated as "men of dark colors." Apart from "dark," *fuscus* also means "brown," "tawny," "dusky," "dim," and "hoarse" (Goodwin, 1874; s.v. "fuscus"). In addition, Blumenbach described the form of the nose as *naso simo*, which should be translated as "flat nose" (Michael, 2017, 295).[8]

---

[5] There are notable exceptions: Farish A. Noor and Peter Carey, eds., *Racial Difference and the Colonial Wars of 19thCentury Southeast Asia* (Amsterdam University Press, 2021); Su Fang Ng, "Making Race in the Early Modern East Indies," *New Literary History* 52, no. 3/4 (Summer/Autumn, 2021): 509–33. https://doi.org/10.1353/nlh.2021.0024. For indigeneity in British Malaya, see Sandra Khor Manickam. *Taming the Wild: Aborigines and Racial Knowledge in Colonial Malaya* (Asian Studies Association of Australia in collaboration with NUS Press, 2015); see also Fenneke Sysling, *Racial Science and Human Diversity in Colonial Indonesia* (NUS Press, 2016).

[6] We have cited here Thomas Bendyshe's translation of Blumenbach's first edition of *De Generis Humani Varietate Nativa*. See Johann Friedrich Blumenbach, *On the Natural Variety of Mankind, ed. 1775*, in *The Anthropological Treatises*, ed. and transl. by Thomas Bendyshe (Longman, Green, Longman, Roberts & Green, 1865), 99-100n4.

[7] For instance, Borneo, Manila, and Ternate are mentioned in Blumenbach's first edition of *De Generis Humani Varietate Nativa* (Vandenhoek: 1776), 91.

[8] Despite this and other possible inaccuracies, we will often quote Bendyshe's translation.

Chronologically, Blumenbach's next publication discussing human variation is his *Manual for natural history*, published in 1779 in German (henceforth: *Manual*). It mentions a fifth human variety, the southerners who inhabited the fifth part of the world. Blumenbach called the peoples of this variety "Australasians" and "Polynesians," and he believed they comprised the populations of the Sundanese islands, the Moluccas, the Philippines, and other islands he did not name ("etc.") or know (Blumenbach, 1779, 64). But the physiognomy of this variety—black-brown skin, wide noses and hairiness—is clearly different from the second variety drafted in Blumenbach's first edition of *De Generis Humani Varietate Nativa* (henceforth: *De Generis 1*), cited above. In terms of geography, the similarity between the *Manual* and the *De Generis 1* is that both works referred to the insular Southeast Asia, Australia, and the islands of the Pacific Ocean. But here we can see an importance difference—the *Manual* omits the continental Southeast Asia previously mentioned in *De Generis 1*. So far, skin color, shape of the nose, and "hairiness" are still the only physiological criteria deployed to define variation in human species; the shape and measurements of the skull Blumenbach adopted later.[9]

Apparently, Blumenbach's view of the Pacific Islands, as well as Australia and New Zealand, as distinct from the mainland Asia, but inclusive of the pen/insular Southeast Asia, was based on the European knowledge on the "new" world geographies increasingly published over the course of the eighteenth century. Use of the terms "Polynesia" and "Australasia" shows that Blumenbach was up to date with these neologisms that circulated in Europe's intellectual circles of his time. These terms were coined by the French intellectual and traveler Charles de Brosses in 1756, the year in which his historical study of travel accounts in the southern hemisphere was published in two volumes. "[T]he fifth part of the world" that Blumenbach deployed in the quotation above "became the norm from the 1770s to the 1830s."[10] In 1776 and 1790, Swedish geographer Daniel Djurberg published a map and accounts in which Polynesia comprised the fifth part of the world—just as it did in Blumenbach's description above. As Douglas writes,

[9] There are numerous studies exploring the ways in which Blumenbach acquired and studied crania or skulls to understand human variation. For an early study on Blumenbach's collaboration with the influential Joseph Banks, see John Gascoigne, *Joseph Banks and the English Enlightenment. Useful Knowledge and Polite Culture* (Cambridge University Press, 1994); John Gascoigne, "Blumenbach, Banks, and the Beginnings of Anthropology in Göttingen, in *Göttingen and the Development of Natural Sciences*, ed. Nicolaas Rupke (Wallstein, 2002); John Gascoigne, "The German Enlightenment and the Pacific," in *The Anthropology of the Enlightenment*, ed. Larry Wolff and Marco Cipolloni (Stanford University Press, 2007); for a recent study on the topic, see Antje Kühnast, "Johann Friedrich Blumenbach's 'Neuholländer'," *Australian Studies Journal* nos. 33 & 34 (2019/2020), 35–37.

[10] Bronwen Douglas, "Naming 'Polynesia': Cartography, Geography, and Toponymy of the 'Fifth Part of the World'," *Journal of Pacific History* 56, no. 4 (2021), 2; 11 (our numbering corresponds with the author approved manuscript available through academia.edu: https://doi.org/10.1080/00223344.2021.1961082.

"[a] line marked 'Boundaries of Asia and Polynesia' runs west of the Philippines and Borneo and through the Sunda Strait, thus positioning all the East Indies except Sumatra with *Polynesien*" (Douglas, 2021, 14–15; original italics).

During his doctoral studies at the University of Göttingen (1772–1775), Blumenbach took a course on the Swedish language, so reading the map would not have been a problem for him.[11] Djurberg's map was translated into German and published in Vienna in 1789, in which the term "Polynesia" was deployed but translated as *Inselwelt* or "island (insular) world." Hence, if not Djurberg's, then analogous maps would possibly have been available to Blumenbach. In 1790, a colleague, friend, and often-quoted expert on the fifth part of the world of Blumenbach's, Georg Forster, published a map based upon his voyage with Captain James Cook in 1771–1775. This map was entitled *Die Inselwelt* ("The Island World") and also omitted Sumatra (Douglas, 2021, 15; map 11 on page 16). Remarkably, Blumenbach's western boundaries of Polynesia excluded Sumatra and started with the Sunda-Straits islands, proceeded eastward to include the Moluccan Islands and the Philippines, and ended in an unspecified "etc." The name for this region was dropped, however.

In his second edition of *De Generis Humani Varietate Nativa*Liber (1781; hencefort: *De Generis 2*), Blumenbach himself explains that he introduced a fifth human variety after having undertaken a "more accurate" investigation of "different nations" in East Asia and America. The five-fold division he now deemed "more consonant to nature" (Blumenbach, 1865, 99n4). The second variety again covered the Asian continent to the east of the River Ganges,"and the part lying beyond the Caspian Sea and the river Obi towards Nova Zembla." This characterization is in direct continuity with the geography for this variety given in *De Generis 1*. In addition, however, Blumenbach gives the following description of the second human variety:

> The inhabitants of this country are distinguished by being of brownish colour, more or less verging to the olive, straight face, narrow eye-lids, and scanty hair. This whole variety may be sub-divided into two races, northern and southern; of which one may embrace China, the Corea [sic], the kingdoms of Tonkin, Pegu, Siam, and Ava, using rather *monosyllabic languages*, and distinguished for *depravity and perfidiousness of spirit and of manners*.[12]

Among the listed northern Asian "nations," or Blumenbach's first "race" of the second human variety (which would indicate at this stage that Blumenbach viewed "race" as categorically inferior to "variety") are the "Manchoos," but also the Japanese. This internal differentiation within the second human variety was a

---

[11] On Blumenbach's studies in Göttingen, and the courses he took, see K. F. H. Marx. "Life of Blumenbach" in *The Anthropological treatises of Blumenbach and Hunter*, ed. and transl. by Thomas Bendyshe (Longman, Green, Longman, Roberts & Green, 1865), 44n1.

[12] Blumenbach (1865), 99-100n4; italics added. In the original, starting from "using other:" "[…] cum linguis magis monosyllabis, tum ingenii et morum nequitia et perfidia." Blumenbach (1781, 51).

novelty, both in relation to *De Generis 1* and to the *Manual.* Corporeal features from *De Generis 1* (1776, 41) and from those ascribed to the second human variety in the *Manual* (1779, 63) match only partially the descriptions given for the same variety in *De Generis 2.*

According to *De Generis 2,* the islands of Southeast Asia and the Pacific, which initially had belonged to the territory of Blumenbach's second variety (the future Mongol "race") grouped together with the Asian continent, were now reclassified as belonging to the fifth variety.

> Finally, the new southern world makes up the fifth [human variety], with which, unless I am mistaken, the Sunda, the Molucca, and the Philippine Islands should be reckoned; the men throughout being of a very deep brown colour, with broad nose, and thick hair.[13]

By reproducing the definition of the fifth human variety from his *Manual,* Blumenbach dropped the names Australasia and Polynesia, and the variety was numbered rather than named - no less due to internal differentiation:

> Those who inhabit the Pacific Archipelago are divided again by John Reinh. Forster [sic] into two tribes. One made up of the Otaheitans, the New Zealanders, and the inhabitants of the Friendly Isles, the Society, Easter Island, and the Marquesas, &c., men of elegant appearance and mild disposition; whereas the others who inhabit New Caledonia, Tanna, and the New Hebrides [Australia], &c., are blacker, more curly, and in disposition more distrustful and ferocious.[14]

Elsewhere, Blumenbach revealed the source of his knowledge on the physiognomy and character of the Otaheitans, New Hollanders, and New Zealanders:

> Every one, for instance, will recognize the fierce and savage countenance of the New-Hollanders and New-Zealanders by looking at the magnificent plates of Parkinson, whereas the Otaheitans, on the contrary, looked at as a whole, seem to be of a milder disposition, as also the many pictures of them by the same well-known author testify.[15]

---

[13] Blumenbach (1865, 100n4). In the original, the inhabitants are described as: "[H]ominibus in uniuersum coloris intense fusci, naso lato, et denso capillo." Blumenbach (1781, 52). Bendyshe translated here *fuscus* as "brown" rather than "dark."

[14] Blumenbach, *Natural Variety 1775*, 100n4, italics added. For some of the present-day names of the mentioned islands, see, for instance, Bronwen Douglas. *Science, Voyages, and Encounters in Oceania, 1511–1850.* (Palgrave Macmillan, 2014). For a detailed study of the two-tribe differentiation of the Pacific Islanders and the aboriginal populations of Australia and New Zealand see Kühnast, "'Neuholländer'."

[15] For the original see Blumenbach (1781, 93). We have used Bendyshe's translation here (Blumenbach, 1865, 123). "Parkinson," whom Blumenbach mentions here refers to Sydney Parkinson who accompanied Captain Cook on his 1771 voyage on the *Endeavour* and who made illustrations of Australian Aboriginals. For Parkinson's biography and a large collection of accessible digitized illustrations, see State Library of New South Wales, "Captain Cook's Voyages of Exploration—Sydney Parkinson," 2005. https://www.sl.nsw.gov.au/stories/captain-cooks-voyages-discovery/sydney-parkinson.

Here, Blumenbach relies on the famous *Observations Made During a Voyage Round the World*, published in German in 1778–80 by Johann Reinhold Forster.[16]

The new variety is called "Malay" in *De Generis 3* (1795, 319–21) for the first time. However, the geographical distribution and internal division of the variety remains the same as in *De Generis 2*. Han Vermeulen thought it was "a mystery why Blumenbach named the fifth variety 'Malayan'," so he offered two possible reasons for this historical choice, which introduced a "linguistic criterion" into the natural history and the study of human variation. One was that Malay was a long-term lingua franca "of the Indonesia archipelago up to New Guinea," while the other was that "[Blumenbach] saw seafaring as characteristic of the South Sea islanders" (Vermeulen, 2015, 373). The linguistic criterion was not an accidental intruder to an already solid(ified) discipline of natural history. Rather, it was one of its "building blocks."

5.3 Language Signifiers of Racial Hierarchies

In *De Generis 2*, the term today known as *Malay* appears in the section in which Blumenbach discusses speech (*loquela*) as one of the unique features of human species. He mentions the renown *orang-utan* (sic) and gives two references in the related footnote (Blumenbach, 1781, 25n*r*). The first of the two is of the early 18th-century *Malaio* dictionary produced by Thomas Bowrey, who decoded *orang-ootan* (sic) as "a Clown, Boor, a wild Man, a Satyr, or Beast found in the Woods of *Borneo*."[17] It is clear here that by reference to this source, Blumenbach could understand the importance of the geographical region circumscribed by the knowledge of Malay language and Islamic trading networks. How Blumenbach discovered Bowrey's Malay-English Dictionary is yet to be determined. Joseph Banks, as

[16] On the career, voyages, life, and publications of John Reinhold Forster and his son George, as well as the impact they had on German Enlightenment scholarship, Blumenbach included, see, for instance, Johann Reinhold Forster, *Observations Made During A Voyage Round the World*, ed. Nicholas Thomas, Harriet Guest, and Michael Dettelbach, with a linguistic appendix by Karl H. Rensch. (University of Hawai ʻi Press, 1996); George Forster, *A Voyage Around the World*, ed. Nicholas Thomas and Oliver Berghof, assisted by Jennifer Newell. 2 Vols. (University of Hawai ʻi Press, 2000); and especially Gascoigne, "The German Enlightenment," and Osterhammel (2018, 184-5; 230).

[17] Thomas Bowrey, *A Dictionary English and Malayo, Malayo and English* (London: Printed by *Sam. Bridge* [sic] for the Author, 1701). The dictionary had introduction to the Malay grammar and dialogues for the learners. It was dedicated to the "Honourable the Directors of the English *East India Company*" and also to the "Honourable the Governour [sic], Deputy Governour [sic] and *Committees* of the Honourable *East India Company*." Its purpose was to promote trade in all the countries where *Malayo* language was spoken, and its publishing was sponsored by the East India Company. The countries where *Malayo* was spoken are listed too on the cover and include "the *Malayo* Country, all the *South-Sea-Islands*, *India*, *Turkey*, *Arabia*, *Morocco*, and generally in all *Mahometan* Countries."

Blumenbach's main British provider of research materials, comes to mind, since "Göttingen was at that time under the British Crown" (Marino, 2022, 2).

We suspect that Blumenbach's genealogical quest for a unified human species with inter-related varieties was inspired by an important scientific duo at Göttingen, namely, Büttner and his colleague Johann Christoph Gatterer, a professor of history.[18] Christian Wilhelm Büttner was Blumenbach's professor of natural history at Göttingen and, according to some, his inspiration for writing a dissertation on human variation.[19] Büttner was a renowned linguist, so he might have obtained a Malay-English dictionary. In fact, Blumenbach (1781, 84) himself describes that Büttner visited London, where he had the opportunity to meet a Chinese person. Apparently, Blumenbach was inspired by Gatterer to think genealogically through his exposure to the study of linguistics at his time at Göttingen, in general, and the categorization of languages according to syllables, in particular.[20] Gatterer had developed his ideas about the inter-relatedness of language and people: if two languages were related, the people who spoke them were too (van Rooy, 2020, 184–5).

Büttner's and Gatterer's interest in the study of Asian languages and/or the deployment of language to study human (pre)history, difference, and kinship ties seems to confirm the unwaning popularity of Gottfried Wilhelm Leibniz's intellectual legacy at Göttingen in Blumenbach's time. Leibniz believed in the common origin of all languages, hence, in the singular origin of humankind, which he looked for in Asia.[21] According to Carhart: "An ancient common language whose traces survived in so many distant vocabularies, Leibniz argued, that the many disparate nations were composed of emigrants or colonists of one single original nation or stock (*gentis aut stirpis*)" (Carhart, 2019, 250). Here we see the usage

[18] One of the leading 18th-century German scholars of the "Göttingen Enlightenment," Gatterer was well informed about the history of Asian nations and was willing to include them in world history. Another Göttingen historian of the period, Arnold Hermann Ludwig Heeren, was of the opinion that the human "race" originated from Asia, which was an ideal place for exploring its history. There is no reason to think that Blumenbach was unfamiliar with Heeren's influential work, which was published in 1793. On Heeren's work see Osterhammel (2018, 9; 51–2).

[19] See "Blumenbach—online.de, Education and Career" https://blumenbach-online.de/Einzelseiten/Biographie-englisch.php#Laufbahn.

[20] Blumenbach enrolled in his doctoral program at Göttingen in 1772, and there is no reason to believe he was unaware of Gatterer's work, as the latter was already an esteemed professor. For these, as well as other topics related to Gatterer and Büttner, see Martin Gierl, "Das Alphabet der Natur und das Alphabet der Kultur im 18. Jahrhundert: Botanik, Diplomatik, Linguistik und Ethnographie nach Carl von Linné, Johann Christoph Gatterer und Christian Wilhelm Büttner," *N. T. M.* 18 (2010), 1-27; Martin Gierl, "Change of Paradigm as a Squabble between Institutions: The Institute of Historical Sciences, the Society of Sciences, and the Separation of Cultural and Natural Sciences in Göttingen in the Second Half of the Eighteenth Century," in *Scholars in Action. The Practice of Knowledge and the Figure of Savant in the 18th Century*, eds., André Holenstein, Hubert Steinke, Martin Stuber, 267-287 (History of Science and Medicine, 2013).

[21] For an extensive study on the topic, see Michael C. Carhart, *Leibniz Discovers Asia. Social Networking in the Republic of Letters* (Johns Hopkins University Press, 2019), 6, 242*ff.*

of botanical *stirpis* taking root in the linguistically approached historical study of the origins of humankind, almost a century before Blumenbach was writing on the (same) topic. The major difference seems to be that Leibniz identified Asia as the continent from which humankind had originated. Blumenbach, in turn, believed that the European "race" was the "primordial" or "original" one. It is likely that Blumenbach was also guided by the argument proposed by Johann Christoph Adelung, regarding Asian languages, hence peoples, as inferior (Metcalf, 1984, 101–15).

The first part of Adelung's *Mithridates* (1806) recorded the translation of the core Christian prayer in various Asian languages.[22] The author elaborated the classification of world languages according to the syllables, introduced, among others, by John Webb, the first to qualify the Chinese language as monosyllabic and primitive (Frodsham, 1964, 389–408). This observation was more than a linguistic remark, though, and served as a "phylo-ontogenetic shortcut," or marker of origin and descendant group (i.e. "race" in modern parlance) in Western scholarship in general, philosophy included. Monosyllabism was equated with the so-called childlike state of linguistic development, and at the moment Blumenbach was reflecting on his racialized varieties, it was considered an indicator of position of the language as a representative of human progress in racial hierarchies (Behr, 2018, 171–3).

Wolfgang Behr notes that it was during the late enlightenment that Adelung turned monosyllabism into a denigrating qualifier for the Chinese people—in opposition to people who speak polysyllabically in the most complex (and hence the most perfect) structured languages, the borders of whose territories were defined practically in the same way as Blumenbach's location of the Caucasian (i.e. European) variety.[23] For Blumenbach, the river Ganges seems to be aligned with the easternmost barrier for Adelung's polysyllabic languages, which ended in Kashmir as the historical cradle of humankind, or Moses' paradise, and whose inhabitants looked nothing like the Mongols or the Chinese (Adelung, 1806, 8–9; 19).

In *De Generis 2*, the peoples of China and mainland Southeast Asia were lumped together in this variety. Blumenbach claimed they all spoke monosyllabic languages and could be "distinguished for depravity and perfidiousness of spirit and of manners" (1781, 51). Adelung classified not only Chinese as a child-like monosyllabic language, but also the languages of the kingdoms on its borders, such as Ava, Pegu, Siam, "Tunkin [Tonkin], Cotschinschina [sic] [Cochinchina],

---

[22] Adelung was following the example set by the Swiss botanist and zoologist, Conrad Genser, who in the mid-sixteenth century published his *Mithridates* containing the translation of the Christian prayer "Our Father" into over 100 then known languages. For a short bibliography of Genser, see George A. Pettit, "Conrad Gesner," *Encyclopaedia Britannica*, last updated December 9, 2024. https://www.britannica.com/biography/Conrad-Gesner.

[23] Behr (2018, 174), who cited the translation to English from Thorsten Roelcke, "Asiatische Sprachen im deutschen Sprachdenken des Barock und der Aufklärung," *Glottotheory* 8, no. 2 (2017), 220. https://doi.org/10.1515/glot-2017-0014. For the original in German, see Adelung (1806, 20–1).

Cambocha [Cambodia]" and Laos (1806, 17; 27). Adelung saw "all these people" as "half-developed humans" who were "greedy," "distrustful" and "incredibly inhumane" towards their enemy during war and retaliation campaigns. He also thought that these people, "some more, others less," resembled those "outstandingly ugly" Mongols with their "flat faces, small and slanted eyes, and snub noses" (32). This characterization reverberated with Blumenbach's understanding of the fifth variety as an earlier stage in human variation.[24]

Blumenbach's description of "spirit and manners" of the people of (in today's understanding) mainland Southeast Asia is not as detailed as Adelung's. Despite being schematic, though, Blumenbach's evaluation does paint a belittling picture of the listed peoples which, in fact, summarizes Adelung's views. As regards physiognomy, apart from the shape of eyes and, perhaps, the facial form, there are no similarities between Adelung's and Blumenbach's descriptions. Nonetheless, in understanding territorial distributions of mutually (un)related Asian peoples, including mainland Southeast Asia, Blumenbach was foremost indebted to Adelung when developing his taxonomy of human variation.

Famously, Blumenbach named his fifth variety in *De Generis 3* "the Malay because the majority of the men of this variety [...] *use Malay idiom*" (Blumenbach, 1865, 275; italics added). To begin, geographical distribution of this variety was further clarified in this edition.[25] Blumenbach included here Pacific islanders, among which were the people of the Sandwich, the Society, and the Friendly Islands, and those from the island of Madagascar, as well as the Eastern Islands, the inhabitants of the Mariana islands, the Philippines, the Moluccas, and the Sundanese islands (1795, 319–20). Except for the Mariana islands, all the other insular entities were listed in the corresponding passage in *De Generis 2*. He informed the reader of *De Generis 3* he relied on the knowledge accumulated by Joseph Banks and Captain Cook on their voyages to the Pacific islands and the southern hemisphere (Blumenbach, 1795, 320n*x*).[26]

---

[24] Viewing races or human varieties in terms of hierarchies organized to reflect the stages of individual human development, in which the lowest intellectual stage corresponds to childhood, is a predecessor of Darwin's evolutionism and a good early example of what came to be known as the stadial theory. Bronwen Douglas relates this theory to Adam Smith and his graduation of societies from "rudeness to civilization" introduced in the second half of the eighteenth century. See Douglas 2014, 110. It is important to keep in mind that stadial theory, or recapitulation theory, as it was also known during the long nineteenth century, was used as a scientific basis for adapting curricula to the "less intelligent" "natives" inhabiting British colonies and U.S. territories (relating to both the indigenous and the African populations) with an emphasis on practical or industrial training and production on readily exploitable and skilled labor. See Thomas D. Fallace, *Race and the Origins of the Progressive Education, 1880–1929* (Teacher's College Press, 2015). We will return to education in our last section.

[25] In fact, as Kühnast notes, Blumenbach explained to Joseph Banks that his hitherto numbered varieties would be turned into toponyms in this edition. Kühnast, (2019/2020, 47). However, as we argue here, this logic was not applied to the Malay variety.

[26] On the works cited in Blumenbach's *x* and *y* footnotes on page 320 and *z* footnote on page 321 see Douglas (2014, 39–40).

An important geographic novelty was Blumenbach's specific mention of the inhabitants of the "peninsula of Malacca and the so-called Indian islands in its vicinity" (Blumenbach, 1795, 295).[27] In comparison, Adelung classified the Malay language as a South Asian language spoken by the inhabitants of a small peninsula he called "Malacca" or "Malaya," and which he believed made up the southernmost part in the backdrop of India, and in the north, directly bordered the monosyllabic Siam. He claimed that Malay appeared to have been a monosyllabic language originally, thus close to the Mongolian and Manchurian languages, but had since evolved into a language *in transition* from a mono- to polysyllabic language. The working hypothesis was that the language had spread throughout the islands of Southeast Asia through migration, which Adelung referred to as colonization, replacing the indigenous languages and peoples or pushing them to the inner parts of islands. Due to the vivid interest in trade, this language had foremost been studied and written about by the English and the Dutch (Adelung, 1806, 100; 102). Among the list of Dutch and British works Adelung quoted was the abovementioned Bowring's dictionary (103). He mentioned the expert on languages, William Marsden, whose Malay dictionary Adelung was eagerly awaiting, intimating that in 1797, Marsden seemed close to completing that task (104). Adelung (589n*) read Marsden's *History of Sumatra* (Marsden, 1784, 2nd edition), but he also read Marsden's early "Remarks on the languages of Sumatera" (Marsden, 1782).

Blumenbach cited Marsden's *History of Sumatra* extensively in *De Generis 3* (1795, 30nu.); 191n*q*.); 193n*w*.); 312n*i*.)), where he also mentioned Marsden's "Remarks" (320n*x*)). The latter were published after Blumenbach's *De Generis 2,* but Joseph Banks was aware of Marsden's hypothesis at the beginning of 1781, if not earlier.[28] Significantly, Marsden believed that "knowledge of the original Siamese, Laos [sic], Cambodian and Peguan languages, as these nations lie in the intermediate space [between continental Asia and Pacific islands?]" would have been of great significance for establishing connections between different peoples, but lamented that he had no such knowledge. Blumenbach learned from Marsden that there was a large language group of inter-related dialects which spread from "Madagascar eastward to the Marquesas, or nearly from the east coast of Africa to the west coast of America [displaying] a manifest connection in many of the words by which the inhabitants of the islands express their simple ideas, and between

[27] However, Bendyshe translates this as "Malayan peninsula" (Blumenbach, 1865, 266). Bendyshe may have reflected the geopolitical reality of his times, when not only Malacca, but other parts of "Malayan peninsula" were under the British colonial rule. Later on, however, he translates the same peninsula as "the Malacca peninsula" (275).

[28] Marsden's "Remarks" was in fact a published letter to Sir Joseph Banks, the President of the Royal Society, and it was read on February 22, 1781. It is not recorded whether the letter was handed to Joseph Banks prior to 1781, which remains a possibility. Marsden (1782, 154). Osterhammel claims that Marsden had been a part of Joseph Banks's intellectual gatherings since 1780. Osterhammel (2018, 195). On the life and career of William Marsden, see the "Marsden's Collection" on the website of the King's College in London. https://www.kcl.ac.uk/marsden-collection.

some of the most distant, a striking affinity," which Blumenbach unified under the name Malay (Marsden, 1782, 155).[29]

5.4 Malays and Native Americans: The Emotionality of Blumenbach's Transitory Varieties

In his final list of varieties in *De Generis 3*, Blumenbach envisaged the Malay variety as in transition between the Caucasian (i.e. European) and the Ethiopian (i.e. sub-Saharan African) varieties (1795, 284–322). Unlike the perceived linguistic unity of his fifth variety, Blumenbach learned from the sources he read that in terms of physiognomy, this variety included great internal differences. For instance, he divided the Otaheitans into two races (sic). The first one included fair and tall persons whose looks were difficult to distinguish from the Europeans. They were the more numerous of the two races.[30] On this variety, Blumenbach (1795, 320ny) quoted Louis-Antoine, Comte de Bougainville, who left a detailed account of the places and peoples he had visited.[31] Then, he proceeded to describe the second race within the Malay variety, drawing on the same source, noting that their size was average or medium and compared their skin color and features with that of "mulattos" (Blumenbach, 1795, 321; cf. Douglas, 2014, 40). The latter race he then approximated to the inhabitants of the southern Pacific islands, who were apparently the third race within the Malay variety

> of whom the inhabitants of the New Hebrides in particular come *sensibly* near the Papuans and New Hollanders, who finally on their part graduate so *insensibly* towards the Ethiopian variety, that, if it was thought convenient, they might not unfairly be classed with them, in that distribution of the varieties we were talking about.[32]

---

[29] Christina Skott argued that although Marsden is usually credited for the concept of the "Polynesian language group," she believed that the idea was presented through travel accounts from Cook's journeys by Forster and Banks. See Christina Skott, "Europe and the Malay World," *Indonesia and the Malay World* 42, no. 123, (2014): 133.

[30] For the original, see Blumenbach (1795, 320).

[31] Louis-Antoine de Bougainville was an eighteenth-century French scholar in the service of King Louis XV sent on a mission across the Pacific to Southeast Asia in search of new colonies for the French East India Company. For visual materials from Bougainville's travels and his short biography and bibliography, see Princeton University Library, "Louise-Antoine de Bougainville, 1729–1811," https://static-prod.lib.princeton.edu/visual_materials/maps/websites/pacific/bougainville/bougainville.html. Although Bougainville's work was translated into English by George Forster and published in 1772, Blumenbach cited here the original published in French in 1771. For Forster's translation, see Lewis de Bougainville, *A Voyage Round the World. Performed by Order of His Most Christian Majesty, In the Years 1766, 1777, 1768, and 1769*, translated into English by John Reinhold Forster (Printed for J. Nourse and T. Davis, 1772).

[32] Here, we cite Bendyshe's translation (Blumenbach, 1865, 275) italics added. We return to the discussion of the italicized terms in the next section.

Blumenbach made this observation following an account by a renowned late sixteenth- early seventeenth-century Spanish navigator of Portuguese descent, Pedro Fernandez de Quirós (1795, 321nz). Blumenbach consulted an English-language translation of de Quirós's account; hence, according to Douglas, he inadvertently replicated the translator's mistake: "This passage does not 'compare' some Islanders to the Ethiopians, as Blumenbach thought, deceived by Darlymple's mistranslation of Quirós's Spanish adjective *loro* as the English noun 'negroes'" (Douglas, 2014, 40). Instead, Douglas has proposed that the adjective *loro* denotes a skin color between white and Black (40n2). The classification of the Malay "race" as a transience between the white and the Black races, based on the perceived "leptorrhinian" quality, circulated among nineteenth- and twentieth-century anthropologists.[33]

The italicized terms in the citation above were intended to highlight the nature of racial transience within the Malay variety, and they testify to emotional labor needed to group certain bodies together with, or away from, other bodies. This (un)grouping was apparently a matter of "sensibility" and was deemed as contingent in nature rather than being based on any objective or scientific fact. Proposed cranial difference was obviously not deemed strong enough to annul the sensibility of approximations based on physiognomy and separate two Black persons divided by territories across the Indian Ocean. Language, in turn, proved to be more reliable—thus, the Ethiopians and the Malays were two different, though related, varieties. The transitoriness as a possibility of approximation or distancing was here to replace *(in)sensible* or contingent separations.

Sara Ahmed has claimed that the "Darwinian model of emotions suggests that emotions are not only 'beneath' but 'behind' the man/human, as a sign of an earlier and more primitive times" (2014, 3). The affective legacy of Blumenbach's varieties was transmitted by means of linguistics, since Blumenbach was academically related to more than one linguist of his time. His relationship with the brothers Alexander and Wilhelm von Humboldt was of great intellectual and cultural significance, as it shows that the emotional labor that went into the making of the Malay was shared by the linguists of the period.[34] Alexander claimed that language was "of the greatest importance in the recognition of similarities or differences in races" (1893, 365). Wilhelm's *On Language* starts with his discussion

[33] For a recent study of the racialized nasal shape in the Malay world, see Nur Dayana Mohamed Ariffin, *Race Manifest. Colonial Administration and Racialisation in Nusantara* (University of Malaya Press, 2022), 36–40. We are grateful to our student Muhammad Nazri Bin Mashud for informing us about this reference.

[34] An abbreviated account of Alexander von Humboldt's life and work can be accessed on Staatsbibliothek zu Berlin, "Alexander von Humboldt Portal." https://humboldt.staatsbibliothek-berlin.de/leben/. In his "Memoir of Blumenbach" Marx mentions that Alexander von Humboldt was among Blumenbach's "pupils." See Marx (1865, 22).

of the "Malayan race." As they populated islands and archipelagos alone, and their language spread "so far and wide," he concluded they were extraordinary navigators.[35]

The admiration for the Malay seafaring skills was an exception and in opposition to mainstream view presented by the stances like Adelung's among European scholars of the period.[36] A belief that Malays pushed the original inhabitants of the insular Southeast Asia, of darker skin, to the inner parts of the islands occupying the coastline in several waves of migration only aggravated this image.[37] The motive of mobility, or impermanence/transience in the parlance of Blumenbach's human varieties, was increasingly used as a negative attribute of the Malay over the course of the nineteenth century.[38]

In the texts by W. Humboldt, the Malayan "race" spread across almost the same regions as those delineated by Blumenbach. The claim that "all these peoples possess such social arrangements that it would be wrong to exclude them entirely from the comity of civilized nations" is a sentimental reflection of Blumenbach's Malay "transience," in line with the stadial theory of the day. Wilhelm reproduced Forster's, Bougainville's and Blumenbach's differentiation into the "almost" Black and white populations within the same variety. Following Marsden, he noted that Black inhabitants occupied the interior of the islands and highlands, while the white populations inhabited coastlines. The darkest and "the most savage" were the inhabitants of New Guinea, New Hebrides, and New Holland (Wilhelm von Humboldt, 1999, 12–13). He was uncertain whether to follow a three-fold or a two-fold internal differentiation of the Malayan "race." Such uncertainties notwithstanding, similar to his brother, Wilhelm insisted on language as a distinguishing, racialized feature:

> It remains, in general, an important question, though on information so far available it can scarcely admit of a satisfactory solution, as to how far older and deeper minglings [sic] of the white and black races may have occasioned gradual transitions in language, and even in colour and growth of hair [...] (Wilhelm von Humboldt, 1999, 15).

Transition in "race" followed/was conditioned by a transition in language, apparently, which meant that they were intrinsically related, if not

[35] Wilhelm von Humboldt, *On Language: On the Diversity of Human Language Construction and its Influence on Mental Development of Human Species*, ed. Michael Losonsky, transl. Peter Heath. (Cambridge University Press, 1999), 11.

[36] For a study on the historicized connection between the Malay and piracy, see Stefan Eklöf Amirell, "The Making of the 'Malay Pirate' in Early Modern European World," *Humanities* 9, no. 91 (2020): 1–14.

[37] For the migration wave theory, See Nur Dayana Mohamed Ariffin (2022, 34–6).

[38] The negative assessment of Malay seafaring skills as a sign of their (inherent) nomadic status was turning into the only legitimate assessment of the Malay in the late eighteenth century. Amirell, (2020, 9). The brothers von Humboldt emotionally belonged to the long eighteenth century, apparently, as their evaluation of Malay seafaring skills was positive.

interchangeable.[39] Among the main sources Wilhelm used for gaining knowledge on Malay language were studies by Marsden and by John Crawfurd.[40] Humboldt continued Blumenbach's classification of Malay and Native American languages as transitory varieties. Such alignment suggests that both were shaped through the same emotional labor in maintaining a racialized hierarchy in which Malay (and Native Americans) held transitory status across the emerging scholarly disciplines.[41] One of those was Friedrich Max Müller's science of language, or what is called "comparative philology" in contemporary parlance.[42] He learned about the Malay language from his professor, Franz Bopp, who, in turn, was a colleague of Wilhelm von Humboldt (McNeely 2011, 140–1).[43] Max Müller rejected the validity of Blumenbach's varieties in linguistics, and attempted to undo the categorical interchangeability between language and race (Max Müller, 1854, 89; 91–2). However, his classification merely reaffirmed Blumenbach's sensibility

[39] Our conclusion here reflects the opinion of Muriel Mirak Weissbach, who noted that Humboldt "rejected the name Polynesian to designate this category [i.e. the Malayan race], on the grounds it was geographical and limited, and preferred to it the term Malaysian, meaning not only the language culture but the people." See Muriel Mirak Weissbach, "Wilhelm von Humboldt's Study of The Kawi language: The Proof of the Existence of the Malayan-Polynesian Language Culture," *Fidelio* 8, no. 1 (Spring, 1999), 36. Foremost, however, the intricate interchangeability of race and language in Wilhelm von Humboldt's work was pointed out by Tomoko Masuzawa, *The Invention of World Religions. Or How European Universalism Was Preserved in the Language of Pluralism* (The University of Chicago Press, 2005), 156–8.

[40] For a comparative study on Marsden's and Crawfurd's ethno-linguistic explorations and theories on the "Malayo-Polynesian" language family, see Martin Müller, "Glotto-Colonial Debates: Language, Nations, Races, and Civilization in William Marsden's and John Crawfurd's Ethno-Linguistics," *Journal of Postcolonial Linguistics* 3 (2020): 1–21. The main methodological difference between ours and Müller's approach is that we focus on a transdisciplinary categorical intertwining of race and language with regard to the Malay, whereas Müller foremost focuses on the disciplinary exchanges between ethnology and linguistics.

[41] Ian F. McNeely's analysis of Wilhelm von Humbold's linguistic methodology and the hierarchization of languages by using "agglutination" and "inflection" as main criteria (instead of the earlier syllables), in which both the Native American and the Malayan languages were ranked as transitional languages is of particular interest here. See Ian F. McNeely, "Wilhelm von Humboldt and the World of Languages," *Ritsumeikan Studies in Languages and Cultures* 23, no. 2 (October 2011): 129–47.

[42] For an extensive study of Max Müller's career and long-term epistemological legacy in the West, see the colossal study by Masuzawa, *World Religions*, in particular chapters 5 and 7. See also Friedrich Max Müller, *Lectures on the Science of Language, Delivered at the Royal institution of Great Britain in April, May, and June, 1861*. 2nd edition. (Charles Scribner, 1862). Project Gutenberg, https://www.gutenberg.org/ebooks/32856, accessed on October 16, 2024.

[43] According to Marlies Spieker-Salazar, Adelung's scholarship "prepared the way for the comparative linguists of the nineteenth century like von Humboldt and Bopp [...] who all contributed to the founding of the Malayo-Polynesian linguistics." See Marlies Spieker-Salazar, "A contribution to Asian Historiography: European studies of Philippines languages from the 17th to the twentieth century," *Archipel* 44 (1992): 187.

as he ranked Malay in the intermediate group of his tripartite division of world languages.[44]

The intermediary position of Malay notwithstanding, the nineteenth-century Filipino anticolonial intellectuals, such as Jose Rizal, have classified Filipinos as Malay in a self-emancipatory claim to the history of a pre-Hispanic civilization in the Philippines. Among the cited authorities were Max Müller and Blumenbach.[45] According to Nur Dayana Mohamed Ariffin, German scholars introduced categories for the study of "race" in the pen/insular Southeast Asia, and influenced the Spanish administrators in shaping their knowledge on Malay and other racialized varieties of people in the nineteenth century (Nur Dayana Mohamed Ariffin, 2022, 7; 24; 52–3; 55; 57). In addition, the German positivist tradition of ethnological and ethnographic studies was adopted in the United States by medical doctors trained in German universities, who, by the "early nineteenth century began to establish their own version of ethnology through an integration of biology and cultural studies." A strong indirect influence on the scholarship concerning "race" was exerted through the presence and work of "German-Austrian scientists [...] in the United States and the Netherlands" (51; 60–1; 66; 72).[46] German landowners were also welcome in the Netherlands Indies in the nineteenth century (Nur Dayana Mohamed Ariffin, 2022, 54–5). In fact, Germans were employed by the Dutch East India Company as experts from the seventeenth century onwards (Osterhammel, 2018, 110).

With such a prominent influence by Blumenbach's race-related studies in the United States, it comes as no surprise that the during their colonization of the Philippines, U.S. colonizers introduced education for the Filipinos by emulating their education of Native Americans. The first General Superintendent for Education in the Philippines, Fred W. Atkinson, had traveled to Germany to complete his PhD, comparing secondary education teachers in Germany and the United States (Atkinson, 1893). Importantly, according to some educationalists, Filipinos were conflated with Native Americans rather than African Americans in the United States, not least because of the former's perceived *assimilability*, which

[44] The ranking of language groups, in line with stadial theory, reflected the purported political maturity, or ability for self-government, of the people who spoke them. Malay was ranked as a "nomad" language, higher than "family" languages but lower than the perfect group of "state" languages. See Max Müller (1854, 22; 24–5).

[45] Apart from Max Müller, in his discussion of the epistemic trajectory leading to Rizal's conflation between Filipinos and Malays, Rommel Curaming also mentions Blumenbach. See Rommel A. Curaming, "Filipinos as Malay," in *Melayu: The Politics, Poetics and Paradoxes of Malayness*, eds Maznah Mohamad and Syed Muhd Khairudin Aljunied, 241–73 (NUS Press, 2011), 246; see also Nur Dayana Mohamed Ariffin (2022, 55).

[46] For the spread of Blumenbach's craniology in the U.S., see Robert J. Richards, "The beautiful skulls of Schiller and the Georgian girl: Quantitative and aesthetic scaling of the races, 1770–1850," in *Johann Friedrich Blumenbach: Race and Natural History, 1750–1850*, eds Nicolaas Rupke and Gerhard Lauer, 142–76 (Routledge, 2019).

the latter lacked.[47] In this sense, the perceived assimilability (into modern citizenship and self-governance) was a reiteration of the familiar Blumenbachian trope of Malay and Native American transitoriness.

Richard O. Winsted, as the British Assistant Director of Education in Malaya, was in charge of shaping the industrial curriculum for the Malay, so in the second decade of the twentieth century he traveled to Java and the Philippines to learn about the basketry making.[48] He saw the Filipino basketry pattern as "more marketable" than the traditional Malay one (Schauer, 2017, 502–3). Ironically, a decade before, an American anthropologist Otis Mason had studied Malay basketry in U.S. museum collections (Mason, 1908, 1909).[49] Basketry and weaving were part of the training in handicrafts introduced from the very beginning at the first higher-educational institution which the British established for Malays in Malaya, more specifically, in Perak, known as Sultan Idris Teacher's Training College (SITC) (Khairul Ghufran Kaspin et al., 2021).[50]

Apart from the official curriculum at the SITC, the Blumenbachian classificatory sensibility of transitoriness, which identified the Malay and Native American "races" as cognates, shaped more informal didactic materials, which currently await further scrutiny. A souvenir book published in 1934 at SITC (see Fig. 5.1) shows a series of Native Americans communicating in sign language with a white North American. The text, entitled "Bahasa Isyarat dalam Negeri Itali dan Amerika" [Sign language in Italy and America] and signed by a certain Abu Hassan Bin Abdullah, provided explanations for a number of illustrations representing the sign language of the "Riadanadian" people of North America (see

---

[47] African Americans, in fact, were seen as inferior to Native Americans, incapable of assimilation and with the prospect of only receiving industrial training for second-class citizenship. See Anne Paulet, "To Change the World: The Use of American Indian Education in the Philippines," *History of Education Quarterly* 47, no. 2 (2007), 185–6.

[48] For Winstedt's trip to the Philippines (and Java), see R. O. Winstedt, "Report by Mr. R. O. Winstedt, Assistant Director of Education, S. S. and F. M. S., on Vernacular and Industrial Education in the Netherlands East Indies and the Philippines," in *Proceedings of the Legislative Council for the Year 1917 with appendix* (Printed at the Government Printing Office, Singapore, by J. E. Tyler, 1918), appendix no. 22; Richard Winstedt. *Start from Alif: Count from One. An Autobiographical Memoir* (Oxford University Press, 1969), 166–82.

[49] For a comprehensive study on influences of Native American basketry on Filipino (Malay) basketry, see Patricia O. Afable, "Crafting Industry: Basket-Making and Early American Education in the Philippines," Mapping Philippine Material Culture. https://philippinestudies.uk/mapping/tours/show/37 (Reprinted from *Back to the Future: Perspectives on Thomasite Legacy to Philippine Education*, edited by Corazon D. Villareal with the original title "Commerce, Indigenous Knowledge, and Museum Science in the Early History of American Industrial Education in the Philippines: the Example of Basket-Making" American Studies Association of the Philippines (2003): 232–255).

[50] This institution of higher education offered agricultural and training in handicrafts using Malay as the language of instruction, and was founded in 1922 in Tanjung Malim, Perak.

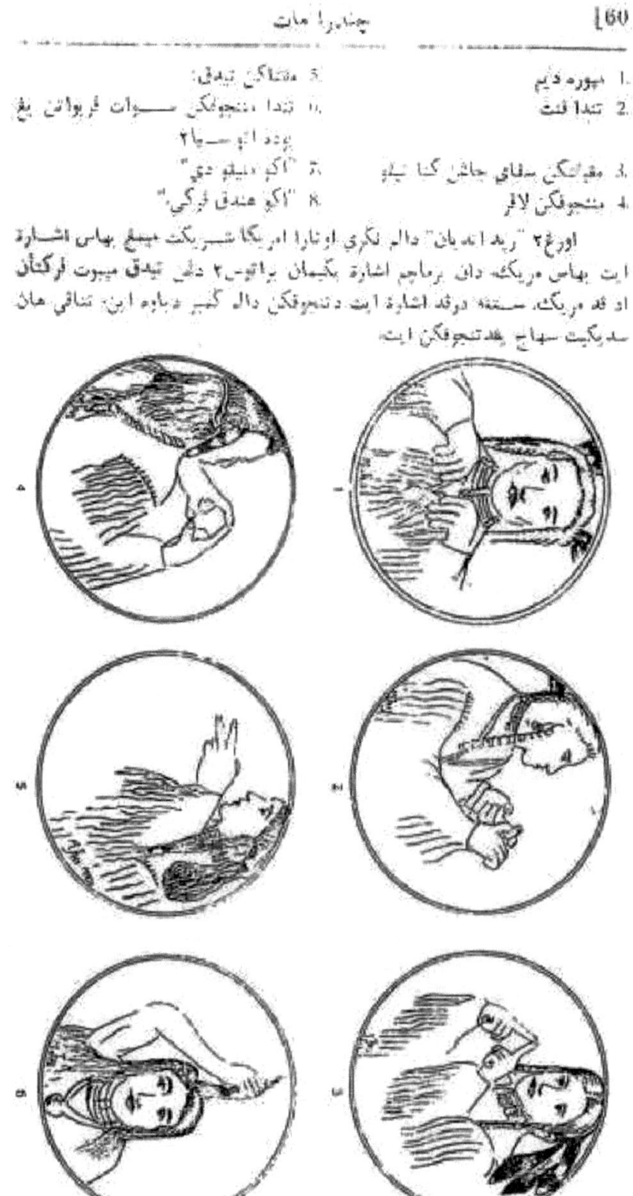

Fig. 5.1 Taken from Abu Hassan Bin Abdullah, "Sign Language in Italy and America," *Chenderamata* 22 (1934), 59–60. Courtesy of Tuanku Bainun Library (Sultan Idris Educational University)

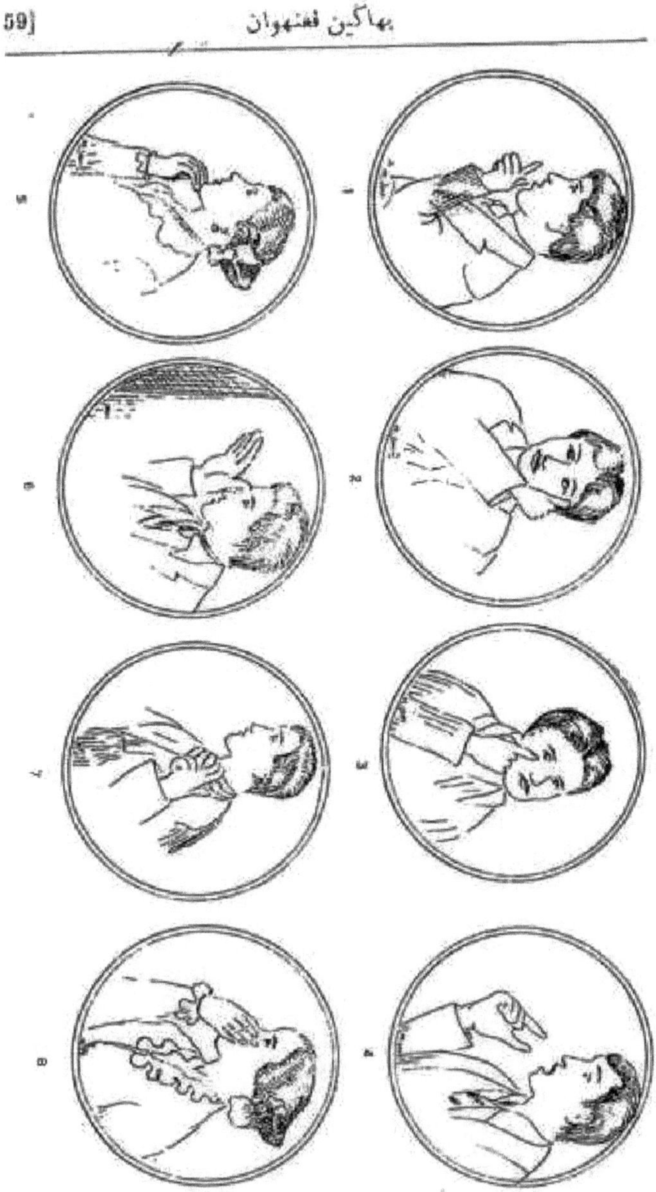

(continued)

Fig. 5.1).[51] Each illustration was followed by an explanation of the meaning of each gesture, and it was also highlighted that the illustrations displayed only a small portion of this language. Abu Hassan Bin Abdullah claimed that according to many a researcher, sign language preceded spoken language, which, in turn, slowly produced written language.

The emotional legacy of aligning two transient "races"—the Malay and the Native American—and the universal domination of Caucasians was clearly informed through the sensitivity that had shaped Blumenbach's categories. Apparently, in the 1930s, and during the Italian occupation of Ethiopia, the position of "extreme" variation fixed at the bottom of the "race pyramid" was abandoned to accommodate white interests, and Ethiopians were "elevated" to the intermediary status of Malays and Native Americans.

Hopefully, future probing into the pedagogical strategies, approaches, and curricula for "natives" across Euro-American colonial empires at the turn of the twentieth century will include the study of the Malay vernacular education in British Malaya in their comparative framework. By approaching educational systems as a colonial governing tool for the dissemination of racial hierarchies, previously understood as seemingly random approximations between unrelated colonized populations, could reveal a classificatory sensibility related to Blumenbach's race science. Comparative research on industrial education in Malaya, India, Africa, and (North) America has yet to be conducted.

5.5 Conclusion

The appearance of the fifth, Malay, variety in Blumenbach's approach to human diversity reflects the complexity of motivations and driving forces behind the crystallization of his explanations regarding the interrelation between human variation and "race." Blumenbach's collective affiliation with numerous colleagues and teachers whose opinion and expert position he shared and respected should be taken into account when recognizing the multiple layers of arguments supporting the idea of transitions between "the most advanced," Caucasian "race" and others. Viewing non-white but also non-Black races as transitory provided multiple options for mixing biological and cultural criteria for multilevel racial hierarchies between and inside of racialized classifications.

Despite the consistent application of linguistic arguments for indicating a fifth (Malay) variation, Blumenbach mostly operated with language without systematic

51 ابو حسن بن عبدالله، بهاس اشارة دالم نݢري ايتالي دان امريکا, چندراماتا: مجله سلطان ادريس ترينيڠ کوليج, ٢٢ (١٩٣٤)، ٥٦–٦١. (Abu Hassan bin Abdullah, ``Bahasa Isyarat dalam Negeri Itali dan Amerika" [Sign language in Italy and America]. *Chenderamata* [Souvenir Book]: *Majalah Sultan Idris Training College* [Sultan Idris Training College Magazine], no. 22 (1934), 56–61). Figure 1 is reproduced on pages 59–60. We are grateful to our student Nurul Balqis Faqihah Binti Mat Asari for romanizing the article for us.

discussion regarding the contrast with other signifiers of human variation. Rather, notwithstanding his ability to bypass binary oppositions, Blumenbach found himself between a hypothesis about the united cradle of humanity and understanding the differences among "races" in terms of their proximity to the "primary," European or Caucasian "race."

In line with this proposal is Terence Keel's argument that race science belongs to the intellectual history of Christianity. With its belief in monogenism, or single origin of human species, Blumenbach and other scientists of his time transformed Christian dogma and a European myth of origin of humans into a basic scholarly premise (Keel, 2018). Blumenbach argued that the Caucasian or European (white) "race" was the "primary" variety from which all others "degenerated." And this both explains the absence of Malay voices in the formulation of their "race" and the need for a historicization that would deconstruct this approach.

We recall here the term *Malayo* which Thomas Bowrey used in his Malay-English dictionary published at the turn of the eighteenth century. There, it is explained that language had a currency not only in the South Sea islands, but also in the Indian Ocean realm (except from East Africa which was not mentioned). The form *Malayo* recalls the Hispanic variant of the term *Malay* which, in the sixteenth century, they recorded as the language spoken in the archipelago today known as the Philippines. As Claudine Salmon (2019) has shown, Malay was the language of seafaring before and at the time of the arrival of the Portuguese in Southeast Asia in the sixteenth century. It was a contact language whose nautical terminology was borrowed beyond the region of the South China Sea, all the way to the southern coasts of China. It was the language learned by the Dutch and used as a means of communication with the Chinese and other ethnicities in the Netherlands Indies until the early nineteenth century. We mention these here as historicized signs of its endurance. As Maznah Mohamad and Syed Muhd Khairudin Aljunied note, the fact that the signification of Malay (*Melayu*) has persisted, can be taken as a sign of ethnic resilience (Maznah Muhamad and Syed Muhd Khairudin Aljunied 2011, xiii).

Our final and not unanimous remarks here reflect the era that also falls outside the timeframe of this chapter—the independence of Malaysia. As Stephen May cautions, "The sociology of nationalism highlights the confluence of nation-building, colonisation, and the related linguistic hierarchization of Indigenous and other minoritized languages" (May, 2023, 654). The Malaysian Constitution establishes Malay as the national language. Whether it is an effect of emancipation or reiterates a historicized colonial link between language and race(ism) will be left for another discussion.

References

ابو حسن بن عبدالله، بهاس اشارة دالم نكري ايتالي دان امريكا, چندرامات: مجله سلطان ادريس ترينيغ كوليج, ٢٢ (١٩٣٤)، ٥٦–٦١.(Abu Hassan bin Abdullah. (1934). “Bahasa Isyarat dalam Negeri Itali dan Amerika” [Sign language in Italy and America]. *Chenderamata* [Souvenir Book]: *Majalah Sultan Idris Training College [Sultan Idris Training College Magazine]*, 22, 56–61).

Adelung, J. C. (1806). *Mithridates, oder allgemeine Sprachenkunde mit dem Vater Unser als Sprachprobe*. 1st part. Vossische Buchhandlung.

Afable, P. O. (2003). *Crafting industry: Basket-making and early American Education in the Philippines*. Mapping Philippine Material Culture. https://philippinestudies.uk/mapping/tours/show/37 (Reprinted from *Back to the Future: Perspectives on Thomasite Legacy to Philippine Education*, edited by Corazon D. Villareal with the original title“Commerce, Indigenous Knowledge, and Museum Science in the Early History of American Industrial Education in the Philippines: the Example of Basket-Making” *American Studies Association of the Philippines*, 232–255).

Ahmed, S. (2014). *The cultural politics of emotion* (2nd edn.) Edinburgh University Press, 2014.

Amirell, S. E. (2020). The Making of the ‘Malay Pirate’ in Early Modern European World. *Humanities, 9*(91), 1–14. https://doi.org/10.3390/h9030091

Ariffin, N. D. M. (2022). *Race manifest. Colonial administration and racialisation in Nusantara.* University of Malaya Press.

Ashcroft, B. (2001). Language and race. *Social Identities, 7*(3), 311–328. https://doi.org/10.1080/13504630120087190

Atkinson, F. W. (1893) *the professional preparation of secondary teachers in the United States.* Breitkopf & Haertel. https://play.google.com/books/reader?id=ulWiAAAAMAAJ&pg=GBS.PP8&hl=en.

Behr, W. (2018). ’Monosyllabism’ and some perennial clichés about the nature, origins and contact of the Chinese language in Europe. In Malinar, A., & Müller, S. (Eds.), *Asia and Europe—Interconnected: Agents, concepts, and things*. Harassowitz, 2018.

Blumenbach, J. F. (1865). *On the natural variety of Mankind, ed. 1775*, in *The Anthropological Treatises* ed. and translated by Thomas Bendyshe. Longman, Green, Longman, Roberts & Green.

Blumenbach, J. F. (1776). *De Generis Humani Varietate Nativa*, 1st edition. Vandenhoek.

Blumenbach, J. F. (1779). *Handbuch der Naturgeschichte.*[Manual for natural history] *Mit Kupfem.* Johann Christian Dietrich.

Blumenbach, J. F. (1781). *De Generis Humani Varietate Nativa Liber.* Vandenhoek.

Blumenbach, J. F. (1795). *De Generis Humani Varietate Nativa* (3rd ed.). Vandehoek and Ruprecht.

de Bougainville, L. (1772). *A Voyage Round the World. Performed by Order of His Most Christian Majesty, In the Years 1766, 1777, 1768, and 1769*, translated into English by John Reinhold Forster. Printed for J. Nourse and T. Davis.

Bowrey, T. (1701). *A dictionary English and Malayo, Malayo and English*. Printed by *Sam. Bridge* [sic] for the Author.

de Brosses, C. (1756). *Histoire des Navigations aux Terres Australes* (Vol. 1 & 2). Chez Durand.

Carhart, M. C. (2019). *Leibniz discovers Asia: Social networking in the republic of letters*. Johns Hopkins University Press.

Curaming, R. A. (2011). Filipinos as Malay. In Mohamad, M., & Aljunied, S. M. K. (Eds.), *Melayu: The politics, poetics and paradoxes of Malayness*. NUS Press.

Douglas, B. (2008). Climate to Crania: science and racialization of human difference. In Douglas, B., & Ballard, C., (Eds.), *Foreign bodies: Oceania and the science of race, 1750–1940.* ANU E Press.

Douglas, B. (2014). *Science, voyages, and encounters in Oceania, 1511–1850*. Palgrave Macmillan.

Douglas, B. (2021). Naming 'Polynesia': Cartography, Geography, and Toponymy of the 'Fifth Part of the World'. *Journal of Pacific History, 56*(4). (Author approved manuscript). https://doi.org/10.1080/00223344.2021.1961082

Fallace, T. D. (2015). *Race and the origins of the progressive education, 1880–1929*. Teacher's College Press.

Ng, S. F. (2021). Making race in the Early Modern East Indies. *New Literary History, 52*(3/4) (Summer/Autumn, 2021), 509–533. https://doi.org/10.1353/nlh.2021.0024

Forster, J. R. (1996). *Observations made during a voyage round the world*, edited by Nicholas Thomas, Harriet Guest, and Michael Dettelbach, with a linguistic appendix by Karl H. Rensch. University of Hawaiʻi Press.

Forster, G. (2000). *A voyage around the world*, edited by Nicholas Thomas and Oliver Berghof, assisted by Jennifer Newell (2 Vols.). University of Hawaiʻi Press.

Frodsham, J. D. (1964). Chinese and primitive language: John Webb's contribution to 17th century Sinology'. *Asian Studies, 2*(3), 389–408.

Gascoigne, J. (1994). *Joseph Banks and the English Enlightenment*. Cambridge University Press.

Gascoigne, J. (2002). Blumenbach, Banks, and the Beginnings of Anthropology in Göttingen. In Rupke, N. (Ed.), *Göttingen and the development of natural sciences*. Wallstein.

Gascoigne, J. (2007). The German Enlightenment and the Pacific. In Wolff, L., & Cipolloni, M., (Eds.), *The anthropology of the enlightenment*. Stanford University Press.

Gierl, M. (2010). Das Alphabet der Natur und das Alphabet der Kultur im 18. Jahrhundert: Botanik, Diplomatik, Linguistik und Ethnographie nach Carl von Linné, Johann Christoph Gatterer und Christian Wilhelm Büttner. *N. T. M., 18*, 1–27. https://doi.org/10.1007/s00048-009-0007-4

Gierl, M. (2013). Change of paradigm as a squabble between institutions: The Institute of historical sciences, the society of sciences, and the separation of cultural and natural sciences in Göttingen in the Second Half of the Eighteenth Century. In Holenstein, A., Steinke, H., & Stuber, M. (Eds.), *Scholars in action. The practice of knowledge and the Figure of Savant in the 18th century* (pp. 267–287). History of Science and Medicine. https://doi.org/10.1163/9789004243910_013

Goodwin, T. (1874). *Latin-English dictionary. Compiled from the best authorities, for use in schools and colleges*. Lockwood & CO. https://upload.wikimedia.org/wikipedia/commons/7/70/Latin-English_dictionary_%28IA_latinenglishdict00good%29.pdf

Government of Malaysia's Official Gateway. "Facts about Malaysia." Last updated on May 1, 2025. https://www.malaysia.gov.my/portal/content/30478

Khairul Ghufran Kaspin, Mohamad Basri Jamal, Norshahrul Marzuki Mohd Nor, Muhammad Hasbi Abdul Rahman, Mazdi Marzuki, Habibah @ Artini Ramlie (2021). "Kurikulum Latihan Perguruan di Sultan Idris Training College (SITC) Sejak Awal Penubuhan Sehingga Perang Dunia Kedua, 1922–1941," *International Journal of Education, Psychology and Counselling, 6*(1), 111–126. https://doi.org/10.35631/IJEPC.639007.

von Humboldt, A. (1893). *Cosmos: A sketch of a physical description of the Universe*, transl. into English by E. C. Otté (Vol. 1). George Bell & Sons.

Keel, T. (2018). *Divine variations: How Christian thought became racial science*. Stanford University Press.

King's College London. "Marsden Collection," May 14, 2020. https://www.kcl.ac.uk/marsden-collection

Kühnast, A. (2019/2020). Johann Friedrich Blumenbach's 'Neuholländer'. *Australian Studies Journal, 33 & 34*, 31–56. https://doi.org/10.35515/zfa/asj.3334/201920.03

Princeton University Library, "Louise-Antoine de Bougainville, 1729–1811," last accessed May 4, 2025. https://static-prod.lib.princeton.edu/visual_materials/maps/websites/pacific/bougainville/bougainville.html

Manickam, S. K. (2009). Common ground: Race and the Colonial Universe in British Malaya. *Journal of Southeast Asian Studies, 40*(3), 593–612. https://doi.org/10.1017/S0022463409990087

Manickam, S. K. (2015). *Taming the Wild: Aborigines and racial knowledge in colonial Malaya.* Asian Studies Association of Australia in collaboration with NUS Press.

Marino, M. (2022). At the roots of racial classification: Theory and iconography in the work and legacy of Johann Friedrich Blumenbach. *BEROSE International Encyclopaedia of the Histories of Anthropology,* 1–28. URL BEROSE: article2546.html.

Marsden, W. (1782). Remarks on the Sumatran languages. *Archeologia, 6* 154–158.

Marsden, W. (1784). *The history of Sumatra* (2nd ed.) London: Printed for the author, and sold by Thomas Payne and Son, et al., 1784.

Marx, K. F. H. (1865). Life of Blumenbach. In *The Anthropological treatises of Blumenbach and Hunter*, edited and translated by Thomas Bendyshe. Longman, Green, Longman, Roberts & Green.

Mason, O. T. (1908). Vocabulary of Malaysian basket work: A study in the W. L. Abbott Collections. *Proceedings of the U.S. National Museum, 35*(1631), 1–51.

Mason, O. T. (1909). Anyam Gila (Mad Weave): A Malaysian type of basketwork. *Proceedings of the U.S. National Museum, 36*, 385–390.

Masuzawa, T. (2005).*The invention of world religions. Or How European Universalism Was Preserved in the Language of Pluralism.* University of Chicago Press

May, S. (2023). Linguistic racism: Origins and implications. *Ethnicities, 23*(5), 651–661. https://doi.org/10.1177/14687968231193072

McNeely, Ian F. "Wilhelm von Humboldt and the World of Languages." *Ritsumeikan Studies in Languages and Cultures* 23, no. 2 (October 2011): 129-47.

Metcalf, G. J. (1984). Adelung discovers the languages of Asia. *Histoire Épistémologie Langage, 6*(2), 101–115. https://www.persee.fr/doc/hel_0750-8069_1984_num_6_2_1192

Michael, J. S. (2017). Nuance lost in translation. Interpretations of J. F. Blumenbach's anthropology in the English Speaking World. *NTM, 25*, 281–309. https://doi.org/10.1007/s00048-017-0173-8

Mirak Weissbach, M. (1999). Wilhelm von Humboldt's study of the Kawi language: The proof of the existence of the Malayan-Polynesian Language Culture. *Fidelio, 8*(1) (Spring, 1999), 29–40.

Mohamad, M., & Aljunied, S. M. K. (2011). Introduction. In Mohamad, M., & Aljunied, S. M. K. (Eds.), *Melayu: Politics, poetics and paradoxes of Malayness.* NUS Press.

Müller, M. (2020). Glotto-colonial debates: Language, nations, races, and civilization in William Marsden's and John Crawfurd's Ethno-Linguistics. *Journal of Postcolonial Linguistics, 3*, 1–21.

Max Müller, F. (1854). *Letter to Chevalier Bunsen on the classification of Turanian languages.* London.

Max Müller, F. (2025). *Lectures on the science of language, delivered at the Royal institution of Great Britain in April, May, and June, 1861* (2nd ed.). Charles Scribner, 1862. Available at: Project Gutenberg, accessed on May 5, 2025. https://www.gutenberg.org/ebooks/32856

Nagata, J. (2011). Boundaries of Malayness: 'We Have Made Malaysia: Now it is time to (re) make the Malays but who interprets the history'? In Mohamad, M., & Aljunied, S. M. K. (Eds.), *Melayu: Politics, poetics and paradoxes of Malayness.* NUS Press, 2011.

Noor, F. A., & Carey, P. (Eds.). (2021). *Racial difference and the colonial wars of 19thCentury Southeast Asia.* Amsterdam University Press.

Osterhammel, J. (2018). *Unfabling the East. The enlightenment's encounter with Asia*, translated by Robert Savage. Princeton University Press.

Paulet, A. (2007). To change the world: The use of American Indian Education in the Philippines. *History of Education Quarterly, 47*(2), 173–202.

Pettit, G. A. (2024). Conrad Gesner. *Encyclopaedia Britannica*, last updated December 9, 2024. https://www.britannica.com/biography/Conrad-Gesner

Povinelli, E. A. (2006). *The empire of love: Toward a theory of intimacy, genealogy, and carnality.* Duke University Press.

Richards, R. J. (2019). The beautiful skulls of Schiller and the Georgian girl: Quantitative and aesthetic scaling of the races, 1770–1850. In Rupke, N., & Lauer, G. (Eds.), *Johann Friedrich Blumenbach: Race and Natural History, 1750–1850*. Routledge, 2019.

Roelcke, T. (2017). Asiatische Sprachen im deutschen Sprachdenken des Barock und der Aufklärung. *Glottotheory, 8*(2), 217–235. https://doi.org/10.1515/glot-2017-0014

van Rooy, R. (2020). *Language or dialect? The history of a conceptual pair*. Oxford University Press (online ed.).

Roque, R. (2014). Race and the mobility of human things. *Science, Technology & Human Values, 39*(4), 607–612. https://doi.org/10.1177/0162243913517146

Salmon, C. (2019). Contact languages on the South China Sea and beyond (15—18th centuries). *Asian Culture, 43*, 20-43.

Schauer, M. J. (2017). Industrious Women and lost traditions: Gender, imperial exchange, and handicrafts education in British Malaya and the American Philippines, 1900–1940. *Journal of World History, 28*(3 & 4), 493–524. https://doi.org/10.1353/jwh.2017.0040

Skott, C. (2014). Europe and the Malay World. *Indonesia and the Malay World, 42*(123), 129–140. https://doi.org/10.1080/13639811.2014.924210

Spieker-Salazar, M. (1992). A contribution to Asian Historiography: European studies of Philippines languages from the 17th to the 20th century. *Archipel, 44*, 183–202. https://doi.org/10.3406/arch.1992.2861

Staatsbibliothek zu Berlin. (2025). Alexander von Humboldt Portal, last accessed May 4, 2025. https://humboldt.staatsbibliothek-berlin.de/work/?lang=en

State Library of New South Wales. "Captain Cook's Voyages of Exploration—Sydney Parkinson," 2025. https://www.sl.nsw.gov.au/stories/captain-cooks-voyages-discovery/sydney-parkinson.

Stoler, A. L. S. (2010). *Carnal knowledge and imperial power: Race and the intimate in the colonial rule* (2nd ed.). University of California Press.

Sysling, F. (2016). *Racial science and human diversity in colonial Indonesia*. NUS Press.

Vermeulen, H. F. (2015). *Before Boas: The genesis of ethnography and ethnology in the German Enlightenment*. University of Nebraska Press.

Visweswaran, K. (2010). *Un/common cultures: Racism and the rearticulation of cultural difference*. Duke University Press.

von Humboldt, W. (1999). *On language: On the diversity of human language construction and its influence on mental development of human species*, edited by Michael Losonsky, translated by Peter Heath. Cambridge University Press.

Winstedt, R. (1969). *Start from Alif: Count from one*. Oxford University Press.

Winstedt, R. O. (1918). Report by Mr. R. O. Winstedt, Assistant Director of Education, S. S. and F. M. S., on vernacular and industrial education in the Netherlands East Indies and the Philippines. In *Proceedings of the Legislative Council for the Year 1917 with appendix*. Printed at the Government Printing Office, Singapore, by J. E. Tyler, 1918. appendix no. 22.

Chapter 6
Pedagogies of White Domination: Blumenbach's Legacy and Racial Classifications in Romanian Science and Secondary Schools, 1870–1914

Cosmin Koszor-Codrea

Abstract The following chapter provides a critical analysis of the introduction and development of racial classifications in secondary schools of the Romanian Kingdom and the Hungarian region of Transylvania during the late nineteenth and early twentieth centuries. In doing so, it traces, on the one hand, the legacies of Johann Friedrich Blumenbach's racial classifications within the broader scientific circles and its further ontological conceptualization as a theory of domination and assimilation of minoritized communities. The chapter further looks at how the development of race science in Romania shifted from theorizing to racially measuring the local Romani and Sudanese ethnic communities. On the other hand, it analyzes the ways in which racial classifications were adopted within the teachings of natural history in secondary schools and how several methodological changes contributed to the legitimization of racial discrimination. The chapter argues that secondary schools, along with other state scientific institutions, are crucial spaces for transforming social realities, shaping in this way a nationalist, anthropocentric and racial hierarchical thinking, while legitimizing at the same time both the domination of minoritized groups and non-human species.

Keywords Scientific racism · Race and ethnicity · Natural history · Blumenbach · History of Romania · History of education

This research is part of broader investigation dealing with the history of Scientific racism and the teaching of natural history in the Romanian and Hungarian secondary schools. The first results of this investigation were published in 2022 in an article entitled "Mismeasuring diversity: popularizing scientific racism in mid-nineteenth century Romania" *Journal for Romanian Studies*, Vol. 4, No. 1 (2022): 37–56. A second part of the results was published in 2023 under the title, "The Tyranny of Schools: Nature and Nation in the Schools of Transylvania and the Romanian Kingdom, 1870–1914" in the *New Europe College Yearbook*. The research was supported by a grant from the Romanian Ministry of Education and Research, CNCS–UEFISCDI, project number PN-III-P 4-PCE-2021-1374, within PNCDI III.

C. Koszor-Codrea (✉)
New Europe College, Bucharest, Romania

V. Shmidt and S. De Angelis (eds.), *Blumenbach's Racial Classification*, History, Philosophy and Theory of the Life Sciences 39,
https://doi.org/10.1007/978-3-032-15950-2_6

6.1 Introduction

Scholarship on the history of scientific racism has pointed to the work of the German anatomist Johann Friedrich Blumenbach (1752–1840) as the main figure responsible for "the shift from a geographic to a hierarchical ordering of human diversity [...] from Linnaean cartography to linear ranking by putative worth" (Gould, 1996: 405). Although Stephen Jay Gould's argument was problematized by several scholars for falsely attributing blame on Blumenbach (Junker, 1998; Rupke, 2019), this chapter argues instead for a more nuanced view on the history of racialized approaches to human diversity and calls for further investigation of several case studies from Central and Eastern Europe. In doing so, the paper will analyze part of Blumenbach's own research on the physical anthropology of the region's minoritized Romani community, the impact of his legacy on various Romanian and international scientists and the adoption of his racial classification theories in secondary schools textbooks published before the interwar period.

Up until now, the German natural science of the Enlightenment and its legacy for Western thought has been the focus of debates among various authors. Studies have ventured to examine Blumenbach's own work and the ways in which his racial classifications have influenced both Christian and secular bodies of knowledge (Keel, 2018: 23–53; Marino, 2022; Raj, 2007: 1308–1309; Rambachan, 2018). Lesser known however, is the adoption and impact of Blumenbach's research on the development of race science in the inter-imperial and semi-peripheral regions of Central and Eastern Europe. This phenomenon is important in many ways, as recent scholarship has highlighted that nation-building agendas in Eastern Europe "were maturing in the context of the so-called 'global reach of whiteness', amid imperial expansion, the growth of race science, and an idea of Europe that was ever more tightly connected to civilizational and racial superiority" (Baker et al., 2024: 16).

On top of this, up to the nineteenth century, the regions of the Black Sea had witnessed a long history of slavery and various forms of unfreedom. In this case, by pointing to both Muslim and Christian persons who lost their freedom, Felicia Roşu has suggested that these forms of unfreedom ranged from domestic and agricultural slavery to other peculiar types of enslavement. However, her study further insists that "the only exception to the regional norm" was Romani enslavement from Moldavia and Wallachia, which "shared several characteristics with North American slavery, such as the stability and self-reproduction of the slave population and the near-equivalence of a certain ethnic origin with the slave status" (Roşu, 2022: 3–6). Moreover, despite various forms of resistance, the oppressive institution of Romani enslavement practiced in the Romanian Principalities and in neighboring regions as well as the local intertwined discussions about national identity, eugenics, and the rise of anti-Semitism gave new meanings to racial classifications. All these discussions led in the interwar period to the deportation of Roma together with the Jewish community during the Holocaust (Petcuţ, 2015; Turda & Furtună, 2021).

Similarly, Victoria Shmidt and Bernadette Nadya Jaworsky, in examining the history of scientific racism in Central Europe, have pointed to the importance of both German and Austrian traditions of physical anthropology for providing theories of assimilation, racial intermixture, the construction of phenotypical binary dichotomies, and building upon Romani racialization. They argue that although during the nineteenth century there were several attempts to theorize Slavic populations as "inferior races" to those in the West, within Slavic countries themselves, Roma continued to be situated at the lowest levels of various racial hierarchies, opposed to other ethnic-religious communities. All these debates, they suggest, have "shaped the racial profile of peripheral Europe and looked at Roma from two interrelated points of view, namely, their proximity to autochthonous populations and their 'true' racial origin" (Shmidt & Jaworsky, 2021: 60–65, 85).

Turning to Blumenbach's own research, he developed his racial theory in his doctoral dissertation *On the Natural Varieties of Mankind/De Generis Humani Varietate Nativa* (1775), while continuing the work of Linnaeus and classifying human varieties into four "types," based on climate, geography, and culture. His thesis underwent several modifications, reaching the third edition in 1795 and adding a fifth variety to his initial taxonomy. In doing so, he classified human "races" based on skull measurement as Caucasian (white), Mongolian (yellow), Ethiopian (black), American (tan-colored) and Malayan (brown) (Baum, 2006: 75–76). Furthermore, when Blumenbach published the fifth edition of his *Manual of Natural History/ Handbuch der Naturgeschichte* in 1797, he was determined to distinguish between two concepts - "race" and "variety," which were initially put forward by Immanuel Kant (Bernasconi, 2006: 84). Likewise, pointing to the beauty of the Caucasian skulls of Georgian women (Figal, 2014: 163–168; Scheibinger, 1990: 387–405) and further identifying the origins of humans in the southern region of Mount Caucasus, Blumenbach's classification postulated that "white" is "the primitive colour of mankind" (Blakey, 1999: 31; Saini, 2019: 36). However, his classification also recognized that the differences between "races" are artificial, while his monogenist view frequently portrayed the unity of humanity, and he expressed an abolitionist stance towards slavery.

Crucially important is the fact that few studies have looked at the entanglements of Blumenbach's theory and Romani racialization, the growth of racial anthropology, and the perpetuation of racial stereotypes within the wider Romanian scientific circles before the interwar period (McMahon, 2009; Piasere, 2019). For instance, after receiving a Roma skull from the Transylvanian based Hungarian physician Sámuel Pataki II (1731–1804) in 1791, Blumenbach not only added what he called the "rarest specimen" of his 64-skull collection,[1] but also reproduced and contributed

[1] Notably, Pataki sent the skull to Blumenbach when his own son, Sámuel Pataki III (1765–1824) was studying medicine in Göttingen since the winter semester of 1787. However, two other Romani skulls from the Russian Empire were received from Georg Thomas von Asch (1729–1807) in 1790 and, after Blumenbach's death, the collection reached a total of five Romani skulls after it received two others in 1830 (including a women's skull) from Transylvania from a certain Alexander von Szentkiraly (see Böker 2019: 3–29). All five skulls are kept today at the Anatomy Centre of the Medical University of Göttingen. I would like to express my gratitude to Wolfgang Böker for further clarifying the development of Blumenbach's collection and to Dr. Nadine Schäfer for providing access to this valuable information.

to the growing trend to criminalize Romani collective identity. Equally important, one year after receiving a Roma skull from the Transylvanian city of Kolozsvár/Klausenburg (now Cluj-Napoca), Blumenbach informed himself of its provenience. In his correspondence from 1792 with Daniel Gottlieb Scheint (1772–1835), by then a medical student in Vienna and later a physician in Mediasch, he discovered the following:

> The gypsy head sent to Blumenbach by [Doctor] Pataki from Klausenburg, is from one of the [sedentary] race and not from wandering horde gypsies, or so-called *Sátor Tziganyok*. In summer, he usually stayed in the woods through which the roads pass and was one of the main thieves and highwaymen of fairs. During one such operation, where he plundered a wealthy peasant, he was attacked by other peasants who were traveling the same road; as soon as they saw his deed, they chased the fugitive robber and caught him. However, the peasants so miserably beat him that he died after traveling several hours to [K]lausenburg and spent two hours in prison. When the corpse was opened, [...] two knives' backs thick was found on the lungs; [...][2] (Quoted in William and Dougherty 2012: 119-220).

All these tragic events occurred a few years after the abolition of Romani enslavement in the Habsburg Empire in 1783, which meant that several persons ended up on the streets, living with no means of existence (Achim, 1998; Furtună, 2020; Petcuț, 2015). However, for the growing field of racial anthropology, these events represented an opportunity to further study the "racial" makeup of various ethnic communities and to establish their origins. Henceforth, Blumenbach provided anatomical analysis of the skull, made additional connections with its social traits, and included an engraving of the skull in the third volume of *Decas altera collectionis suae craniorum diversarum gentium illustrate* (1793). Moreover, he reinforced various stereotypes of Roma's "irregular and savage way of life," living as "vagabonds," arguing that besides their involvement in various robberies they lured children, while their own offspring were the result of diverse combinations of parents. He also insisted that if the local Transylvanians would have seen his skull collection, they would be able to discern a Romani skull. Contradicting H. M. G. Grellmann's (1756–1804) theory that situated Romani origins in Hindustan, he implied that based on the skull anatomy and its resemblance with another mummy skull from his collection, Roma had ancient Egyptian origins (Blumenbach, 1793: 3–5; Willems, 1997). In doing so, Blumenbach labeled the skull "Cingari Genuini," which he further placed within the "Intercaucassiam et Etiopicam" classification group, thus establishing a future trend in physical anthropology that debated, Roma's so-called "racial character" (Blumenbach, 1795: XXIX; see also Ruch, 1986: 159–166). Finally, by 1795, while reflecting on Romani and Jewish "racial character" from Transylvania and making comparisons with Germans, Blumenbach again built on their "otherness" by arguing that their configuration was shaped by the lack of intermarriages with other ethnicities:

---

[2] The letter written by Johannis Binder also revealed that Daniel Scheint had prepared the skull for collection purposes and had been forced "to leave his old quarters, because the landlord thought that the house had been haunted ever since". (William and Dougherty, 2012, 220) At this stage in my research, I neither located further information on these tragic events nor discovered the identity of the person studied by Pataki and Scheint.

> The ancient Germans gave formerly instances of the unadulterated countenance of nations unaffected by any union with any other nation, and today the genuine Zingari, inhabitants of Transylvania do the same; and above all the nation of the Jews, who, under every climate, remain the same as far as the fundamental configuration of face goes, remarkable for a racial character almost universal, which can be distinguished at the first glance even by those little skilled in physiognomy, although it is difficult to limit and express by words (Blumenbach, 1795: 223-234).

Regarding the institutions that trained the wider populace on how to read people through racial anthropometric lenses, some can be identified within the local public school system. Although several studies have analyzed the development of secondary education in Romania and in the Austrian-Hungarian Empire (Berecz, 2013; Caramelea, 2015; Puttkamer, 2003; Ungureanu, 2015), there is little research that deals with the ways in which schools were agents of perpetuating racial thinking in Central and East European regions. For instance, the teaching of natural history in the Romanian schools of Transylvania dates back to the late eighteenth century, when the Austrian Empress Maria Theresa (1717–1780), through the act of *Ratio Educationis* from 1777, mandated that the study of nature in secondary schools was to be divided according to the three natural kingdoms: zoology, botany, and mineralogy (Chețianu, 1902: VIII). Notably, when it came to the teaching of zoology, children were introduced to anthropocentric views of nature and were instructed through simplified taxonomies in the hierarchical study of human racial classifications following Blumenbach's work. Further, up to the beginning of the twentieth century, several methodological changes in the teaching of natural history, following school reformers such as August Lüben (1804–1874), Pál Gönczy (1817–1892), and Albert Apponyi (1846–1933), were not only subverting education to the Hungarian nation-state political agenda, but at the same time, some of the most famous racial theories of the period were mandatory knowledge for all natural history pedagogues.

Looking at the impact of official scientific narratives on secondary education, Ann Morning has shown that schools have always played an important role in the maintenance and preservation of racial discussions put forward both by naturalists and lay people. While analyzing the development of school biology textbooks in the United States, Morning argues that "textbooks offer an important vantage point on the confluence of scientific and popular concepts of race. They reflect—albeit imperfectly—the scientific race views of their era, and they reveal what messages about race are being widely presented to the public as 'scientific.'" Morning further stresses that "textbooks embody a hybrid knowledge, informed by both expert and lay imagery. We can read them as results of a process of 'working out' scientific ideas for public consumption and credence. As such, they offer insight into how certain racial beliefs are fashioned, made credible, and perpetuated" (Morning, 2008: S.111; see also Brosnan, 2016: 718–733). Similarly, Stephen Heathorn has shown that within the English educational system, while various textbooks were praising national identity and imperialism, school children also learned that "concentric circles of racial belonging and identification were present in these books ranging from the 'races of mankind' based simply on colour to considerations of the subracial groupings within each colour" (Heathorn, 2000: 94–95). Hence, racial classifications circulating in school textbooks were an integral part of the wider interests in development of race science in the nineteenth century, strengthening racial stereotypes and building on representations of white identities.

Based on the above-mentioned scholarship, there are clearly several gaps in understanding the role played by Eastern European scientists in the production of a scientific image of humans as divided into races and how they affected minoritized communities. The major channels for disseminating knowledge about racial classification and hierarchical thinking, namely secondary schools, remain underexamined. Therefore, this chapter will first examine the impact of Blumenbach on Romanian naturalists and how racial classifications were first adopted within local scientific circles. It further analyzes the adaptation of race classification theories to ethnic groups and highlights the range of analogies and comparisons introduced by naturalists, which served racist prejudices against minoritized Romani and Sudanese communities living in the Romanian Kingdom. Secondly, the chapter will demonstrate the ways in which Blumenbach's concept of a "white Caucasian race" remained important for those doing fieldwork in racial anthropology. It highlights how racially informed research provided answers to questions about Romanian (white) national identity. The final section explores the ways in which racial classifications influenced secondary schools and how these institutions further shaped nationalist, anthropocentric, and racial hierarchical thinking, while legitimizing both the domination of minoritized communities and non-human species through racial language.

6.2 Blumenbach's Legacy for Romanian and International Scientists

During the early nineteenth century, the first generation of naturalists to promote racial classification theories within the Romanian Principalities of Moldavia and Wallachia were physicians trained in German academic centers such as Heidelberg, Leipzig, and Berlin. For them, the adaptation of Blumenbach into the Romanian language was aimed at remedying a lack of scientific literature, as well as spreading racial classification theories and creationist ideas about the natural environment. A second generation of native Romanian scholars trained in Paris put forward their ideas soon after the creation of the modern Romanian state around the 1870s. For them, racial theories went beyond their mere adaptation to the Romanian language and overlapped with hierarchical racial ideas of Romani assimilation and narratives about Romanians' own origins and "superior" national identity. Similarly, after the institutionalization of physical anthropology in France and Germany, between the 1870s and 1914, international scientific networks brought physical anthropologists such as Izydor Kopernicki (1825–1891) and Eugène Pittard (1867–1962) to Romania, who were also eager to contribute to local debates on scientific racism. Their field research was in line with Romanian and French views on racial anthropology, giving new meanings to the idea of race by focusing on understudied ethnic groups. However, although the two German and French schools had conflicting scientific ideas in terms of methodological approaches and often had contradictory research outcomes, both generations were influenced on the one hand by Blumenbach's identification of the so-called "Genuine Gypsy" and, on the other hand, deployed sophisticated skull measurements and further built on the racialization of Roma (Piasere, 2019: 105–106). Finally, by the interwar period, Romanian racial anthropologists and serologists, especially those working in Transylvania, had turned towards the German school, which by now "offered a tightly integrated complex of right-wing völkisch nationalism, an obsession with ethno-racial purity, Nordic supremacism and the anthropological sub-discipline of raciology" (McMahon, 2009: 102).

As part of the first generation of physicians interested in physical anthropology, the spread of Blumenbach's racial classifications in the Romania Principalities of Wallachia and Moldavia dates back to the first part of nineteenth century. In 1837, German-born physician living in Iași, Iacob Czihak (1800–1888), published the first scientific textbook in the Romanian language entitled *Manual of Natural History*. Appealing to Blumenbach's thesis and slightly changing the classification order, Czihac's adaptation from first hand textual sources described in detail the differences between the human "races," insisting on the bestialization of non-Europeans. Hence, Czihac's classification mingled with comparisons between Ethiopians and Malayans, and the affinities of non-human species such as monkeys, attributing to them an extreme state of savagery; the only exceptions within the "black/white polarity" were Mongolians and Native Americans, which as Renato Mazzolini argued, were best seen as arbitrary abstraction (Mazzolini, 2014: 144). In Czihac's words:

> 1. The Caucasian race has the most regular beautiful body form and white skin color [etc.]. This race is comprised by Europeans (without the Finnish and Laps), including some Asians, Arabs and Tatars. 2. The Mongolian race with flat face and forward projections of their cheeks, are the most numerous. 3. The Ethiopian race or Negroes, more or less black, with woolly hair, prominent jaws, thick lips, flat nose and cannot sneeze. Given their shape, they resemble monkeys more [than humans]. 4. Malayan race, dark skin, flat nose, big mouth, dense black and curly hair […]. Some of these people are extremely savage, ugly, and they eat humans, while others are cultivated. 5. American race, copper or cinnamon colored, wide face, flattened, thin black hair and rarely bearded (Czihac, 1837: 16-17).

Czihac published his work at a time when, throughout Europe, the number of handbooks of natural history dealing with the explanation of skin pigmentation of Saharan Africans, reached the highest peak (Mazzolini, 2014: 137–138), while the global invention of the idea of "whiteness" was also under consideration (Kolchin, 2002: 154–173). Likewise, around the 1850s, other physicians and natural history schoolteachers working in Bucharest, including the German-trained Jewish physician, Iuliu Barasch (1814–1863), and Dimitrie Ananescu (1831–1885), also promoted the spread of Blumenbach's idea of a "white Caucasian race" in their science popularization journals, secondary school manuals of natural history, and public lectures. (Koszor-Codrea, 2022: 37–56)

A second generation of Romanian scientists trained in Paris and interested in racial classifications put forward their arguments around the 1870s. For them, the adaptation of racial theories in Romania was not just another starting point in the enumeration of racial thinkers, but overlapped with reference to the Romanian's own mythological ancient origins and racial character. For instance, one of the first research studies of Romanians in physical anthropology and craniology was led by Mihail G. Obedenaru (1839–1885), who, in 1874, presented his three skull case studies to Société d'Antropologie de Paris, of which one he claimed was "Dacian," as it "resembled the Dacian figures represented on the Trajan Column" in Rome (Turda, 2007: 361–377). By the use of distorted scientific evidence, he was particularly interested in associating the Romanian nation with what the French physical anthropologist Paul Topinard renamed the Celto-Slav race, and by doing so, it upgraded them to the status of superior racial Aryans and linked them with the civilization of France (McMahon, 2019: 28–29).

Another scientist who carried out research on craniometry in Romania was the Polish physician Izydor Kopernicki (1825–1891). Kopernicki crisscrossed international networks of physical anthropology; he was a member of the Paris and London Anthropological Societies, and honorary member of the Society of Galician Physicians in Lwów (now Lviv). After being exiled for nationalist agitation, he ended up working in Bucharest at Colțea Hospital and writing for the German *Archiv für Anthropologie* he, presented himself as a homeless European researcher. As Maria Rhode highlighted, Kopernicki's remarks on skin pigmentation were influenced by Blumenbach's own research and adopted the French chromatic scale. Referring to traditional categories of social classification, he argued:

> As 'among the inhabitants of our land, skin does not have particular colors and differs only in the grade of the suntan of white skin, three terms should be sufficient for description, white, beautiful white without any suntan, buff, similar to the usual skin of our peasants, swarthy, […] like the so called Eastern skin, common for Jews and Armenians and most typical for gypsies' (Quoted in Rhode, 2019: 115).

Moreover, in his study from 1872, on the conformation of Romani skulls from Romania, not only did Kopernicki begun his argumentation by reproducing Blumenbach's description of the "Cingari Genuini" from 1793, but he also set out to analyze both the social and "racial" characters of 15 male and 5 female Romani skulls. In his introduction, Kopernicki opined that Roma were "generally known as clever, skillful, smart and sufficiently intelligent people"; however, he further identified the reasons they were "unreceptive to our cultural influences." Similar to previous research on the topic, the answer he gave had to do with the "unconquerable inclination towards a carefree, duty-free, roaming about and the shyness of a hard cultural life, is what makes the Gypsies the way they live and will long remain as they have been known everywhere and at all times: a vagabond race" (Kopernicki, 1872: 268–273). Turning to the scientific aspects of his measurements, Kopernicki highlighted that based on the "volume and capacity of gypsy skulls" this showed they were "inferior to all European skulls, even to the most brachycephalic ones." In addition, while contradicting Austrian racial anthropological research led by Augustin Weisbach (1837–1914), he insisted that the examined skulls were "orthocephalic, with some tendency to dolichocephaly," which positioned them closer to the so-called "lower races." He further concluded that according to "the position of the occipital foramen which was very similar to that of the Negro skulls, this shifted more towards the centre in the Gypsy skulls than in the white." Finally, his racialized views went hand in hand with gendered categories ascribed to Romani women; He argued that "the female gipsy skulls presented a much greater degree of diminution in their construction than the male" (Kopernicki, 1872: 283–303).

The history of Romani community reveals that, although the abolition of slavery officially ended in the Romanian Principalities in 1856, after a period of 500 years, the Romanian authorities financially compensated former slave owners. Historian Petre Petcuţ has shown that this process left those people who refused to work on the oppressive boyar's land and households on the streets (Petcuţ, 2015). Within this context, around the 1870s, one of the main concerns of Romanians in the natural sciences, such as Ion Ghica (1816–1897), by then the president of the Romanian Academy, was how to deal with the newly freed Roma in scientific and racial terms. Ghica was a natural science graduate from Paris and remained passionate about the works of Georges Cuvier (1769–1832), Elie de Beaumont (1798–1874), and Alexander von Humboldt (1769–1859) (Ghica, 1967). In terms of publishing, he began writing popular articles as early as 1844 dealing with the specialisation of various scientific disciplines. Soon afterwards, he put forward his curiosity in archaeology and humans' place in nature, and mingled his interests with racial classification in several studies written between 1866 and 1884: *Human and intellect*; *The Beginnings of Humans; Stirpe, Species or Races*; *Humans and the Earth* (Istrati, 1902: 16).

In the paper, *The Beginnings of Humans*, delivered at the Romanian Academy in 1878, Ghica shifted between various creationist explanations of the origins of the earth and the geographical distribution of humans. His study avoided addressing Darwin's question of human evolution, arguing that the "state of science" had not fully reached a consensus on the issue. He, instead advocated, in accordance with Georges Cuvier, for the theory of catastrophism (Ghica, 1878: 61–63). In the second part of his lecture, dealing with racial classification, his ambivalence was revealed. First, he promoted the monogenist view of Humboldt's "unity of mankind" by arguing, "The consequence of his theory is that the civilised populations are not distinct from savages or troglodytes." However, his presentation took a radical turn after he systematically summarized that races are best grouped into Caucasian, Mongolic and Ethiopic, and insisted that "white people […], pure or mixed with other races, are without doubt the most intelligent, giving birth to the most civilised nations" (Ghica, 1878: 74–75). Supporting his statement with references to Blumenbach's work and scholarship produced by a plethora of other French racial anthropologists, famous among the Romanian late nineteenth century physicians (Bărbulescu, 2014; Turda, 2024), Ghica's views on the intersection of gender and racial dynamics of power went as far as to claim that whiteness remained extremely important even after the abolition of slavery in the United States:

> Women of colour, whether red or black, have an appeal to white men; the dark colour is considered as evidence of inferiority. Blacks feel humiliated by the colour of their skin; the greatest offence to them is to tell them that they are black; after the abolition of slavery in the South of the United States, black [people], instead of calling themselves free, they say that they are white (Ghica, 1878: 84).

Equally important was Ghica's continuing ambivalence as evidence by the communication delivered at the Romanian Academy in 1880, suggestively entitled, *Stirpes, Species and Races.* On this occasion, he considered that racial classification theories might be applied to the local Romani community, further drawing on South American colonial practice:

> Mr. Broca and other anthropologists have observed that, when civilised, the third generation of the Lobi tribes in South America acquired a different appearance to those who remained savage. The same can be said about our urban gypsies, who although they have the same bond with those from Dobreni, the laie [i.e. nomadic Roma] and netoţi [i.e. nomadic Roma], they differ in their appearance. Hence, the netoţi established in Vizureşti have totally changed their physiognomy, after their settlement (Ghica 1973: 106).

Ghica's departure from the Humboldtian view of the unity and harmony among species was irreversible, as his argument continued to oppose, together with Blumenbach and Arthur de Gobineau, interracial mixing, while also putting forward his views on race and national identity. For instance, he claimed that there was "a natural instinct which pushes humans, as well as animals, to preserve the purity of their descendants," and that "preserving the purity of a nation ensures its existence" (Ghica 1973: 116–117). Equally important was the fact that his statement also overlapped with the enactment of the 1881 "foreigners law," directly mandating not only the expulsion of local anarchists, but also Jewish and nomad Roma traveling around the country. As Petre Petcuţ has shown, the law prohibited the latter to settle anywhere without proof of income and without the acceptance of a rural or urban committee. Furthermore, if they failed to provide a travel document, they were expelled on the basis that their mobility and activities was now considered a criminal offence, "vagabondage" (Petcuţ, 2015: 156–171).

6.3 From Racial Theorizing to Scientific Practice

At turn of the nineteenth century, a new generation of Romanian racial scientists, in collaboration with internationally recognized anthropologists, shifted from theorizing to racially measuring existing multi-ethnic communities. In doing so, their obsession with "purifying the nation," also switched to the recently acquired territory of Dobrudja (Iordachi, 2017: 291–340). The complexity of this multi-ethnic semi-peripheral region near the Black Sea led to new problems. Several ethnic-religious communities inhabited the land, including Tatars, Kurds, Muslim Roma, Circassians, Jews, Russians, Romanians, Lipovans, Skoptsy, Laz, and, the Sudanese; the latter were highly debated by the period's racial anthropologists. The Sudanese, a community of 145 families, were also enslaved by the Ottoman Empire and were colonized in the 1830s in the region of Dokuz Agaç (now Măgura), as well as in other five villages for their agriculture skills in dry climates (Ionescu, 1904: 343; Hathaway, 2018). As the region became part of the Romanian Kingdom, several attempts at colonization brought both scientific and religious "white expertise". Hence, the Romanian Greek-Catholic priest born in Transylvania (part of the Hungarian rule of the Austrian-Hungarian Empire), Vicențiu Grozescu (1838–1897), after visiting Constanța, pleaded "for the establishment of order and the spread of Romanian civilization" while portraying the Dobrudjean Muslim community in racial terms. Traveling to the market of Medgidia, he not only made anti-Semitic observations related to the local Jewish community, but when encountering the Sudanese, he described them as "Arabs black as ink, with sneered teeth and lips as those of horses that grow old in Opler's beer houses." His racial views overlapped with his anthropocentric observations and were contrasted to the so-called "vigorous Romanian peasant from Transylvania." He insisted on Orientalising tropes that depicted Dobrudja's regional culture as mired in a "primitive" and "animalistic" state of evolution (Grozescu, 1885: 321–322).

Similarly, when the same Sudanese community was discussed in 1907 in one of the most important Romanian literary journals, *Convorbiri Literare* (*Literary Talks*), the article in the anthropology section argued, "Today, in most countries, you have the opportunity to see people of different races. Dobrogea, conquered from the Turks, brought under our rule, besides the Tartars, even some blacks from Sudan." Moreover, after exclaiming, "Who would believe that black people exist until he sees a black man!" the article pointed to the importance of Blumenbach's research by arguing that "it's only been 100 years since the colour of human skin became better known. [Hence], Blumenbach was the first to divide humans into white, black, yellow, red and olive" ("Cronica Literară și științifică: Fapte" 1907: 110–111).

During the same period, the region of Dobrudja was also researched by the Swiss racial anthropologist Eugène Pittard (1867–1962). At the beginning of his career, in 1899, Pittard was granted assistance for and permission to carry out his first anthropological measurements of the Romanian multi-ethnic territories (Turda, 2018: 40, 110; McMahon, 2009). His findings were eventually published in several issues of the *Bulletin of the Science Society* (*Buletinul Societății de Științe*), from Bucharest. Meanwhile, Pittard's physical anthropological research, carried out in 1901 on Romani communities of Dobrogea, Râmnicu Sărat, and the Prahova Valley, was presented at the Romanian Association for the Advancement and Spread of Science congress held in 1903. At that time, he registered no more than 1300 individuals and divided them based on national and ethnic origins ("Romanian, Turkish and Bulgarian Gypsies"). A total of 837 were men and around 500 were women, and Pittard sought patterns of interbreeding (Pittard 1908: 436). His conclusions were in line with the Romanian congress's assimilationist view of Roma, winning him a golden medal for his "scientific exposé" and opening new opportunities for future research in the region. Moreover, Pittard also reached the Sudanese community, measured their body features, and claimed that although there were signs of "interbreeding" with other ethnicities (Fig. 6.1), they embodied a so-called "negroid character" (Pittard, 1913a: 147–162; Pittard, 1913b: 307–328). Finally, investigating Pittard's scientific contributions and his tenure as a president of the Gypsy Lore Society, Thomas Acton has highlighted that Pittard's work was a "major conduit of scientific racism," crucial for strengthening a racialized view of Roma as "primitive outlaws." Acton insisted that:

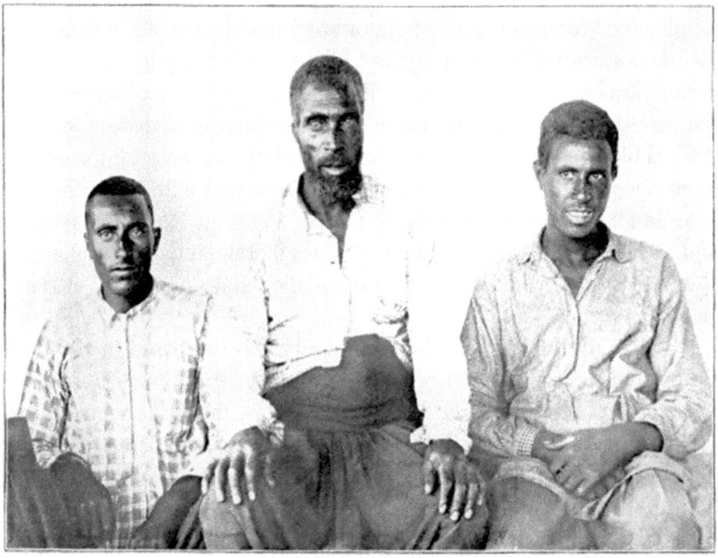

Le Globe, LII, Mémoires — E. Pittard. Contribution à l'étude anthropologique des Arabes — Pl. I

Arabes (de la Dobroudja). L'homme qui est au milieu a un peu de sang nègre. L'implantation des cheveux n'est pas celle des Arabes.

Le Globe, LII, Mémoires — E. Pittard. Contribution à l'étude anthropologique des Arabes — Pl. II

Les mêmes que dans la figure 1. On remarquera que l'individu métissé (au milieu) n'a pas le nez caractéristique des Nègres. Il est aussi beaucoup plus barbu que ceux-ci.

Fig. 6.1 Sudanese peasants from Dobrudja photographed by racial anthropologist Eugène Pittard Source Pittard Eugène, Contribution à l'étude anthropologique des Arabes. In: *Le Globe. Revue genevoise de géographie*, tome 52, 1913. pp. 147–162 Courtesy: Sig.: I-383393/52.1913, Universitätsbibliothek Wien

> Pittard's central argument was ultimately circular: he tried to identify the core original Gypsy physical anthropology by excluding those who were not 'vrais tsiganes', ('true Gypsies') but ultimately, in a different intellectual climate, had to concede that nomadism was not a preserver of racial purity, and that the diversity of the data might indicate a complex origin (Acton, 2015, 5-7).

Hence, Acton concluded that "the peculiar nature of that racism, and the paradoxical ideology of 'The True Gypsy', [contributed] to the oppression and genocide of actual Roma/Gypsies/Travellers" (Acton, 2015, 5–7).

6.4 Representations of "Whiteness" and Racial Classification in Schools

Another important aspect needing further exploration includes the ways in which the complex and contradictory research in racial anthropology was simplified and translated into educational and instructive messages in the textbooks of secondary schools, as well as how views of racial superiority and racial slurs were perpetuated among schoolchildren. Scientific ideas about race circulated widely in the public sphere during the nineteenth and early twentieth centuries in the schools of the Romanian Kingdom and the Austrian-Hungarian Empire. These ideas were negotiated and contested by different conflicting ethnic groups, each aiming to build upon and protect their own national, religious, and political status. As Marius Turda has clearly shown, before the interwar period, fears of race degeneration and ideas of national superiority, led Hungarian physicians, zoologists, and politicians to place Social Darwinism, race, and eugenics at the heart of the Hungarian state political agenda. Their aim was to "purify the national body," in order to reach its so-called historic destiny in the Carpathian basin (Turda, 2022). In response to these politicized claims, Romanian intellectuals, while refuting Hungarian Magyarization and assimilationist plans, built their own racial doctrines of national survival. For instance, the Romanian schoolteacher from the Transylvanian city of Năsăud, Vasile Meruțiu (1881–1943), argued:

> It is not condemned that the Romanians from the Hungarian state learn in their schools also the Hungarian language. […] As a proof there are 500 young Romanians learning in high superior Hungarian schools. […] They learn in Hungarian language with much effort, however in their chest is boiling the racial hatred, which is pushed by the intolerance of Hungarian chauvinism (Meruțiu 1906: 24).

Similarly, other racial and anthropocentric articles published in the Romanian newspaper from Hermanstadt (now Sibiu) concerning the struggle of national recognition argued that "the political term "race" used to refer to the non-Hungarian nationalities, can be applied if you will pardon the expression, to the race of horses, not to people." However, after pointing out that the concept of "race" existed only in relation to European colonial expansion, the author argued, in accordance with Blumenbach, "We Romanians will always confess that we are part of the Caucasian race and together with the French, Italians, Spaniards and Portuguese form the queen-kingdom called Latin. And as a people we call ourselves with great pride: Romanians" (I.M. 1889: 237–238). However, as Anca Pârvulescu and Manuela Boatcă have clearly shown, in inter-imperial regions such as Transylvania, "Romanians' claim to freedom, in particular, is anchored in the Roma's position below them." Moreover, both authors have critically highlighted that "Romanians are Romanian not because they are not Hungarian—the most common dichotomy in the context of Transylvania within the Austro-Hungarian Empire—but because they are not Roma" (Pârvulescu & Boatcă, 2022: 84). By doing so, the success of the Romanian "ethnic project of becoming white" (Treitler, 2013) depended not only on their racial distancing from the local Romani community, but also on promoting scientific ideas about their own presumed "white Caucasian" identity, including racial theories that were part of the secondary school natural history curriculum.

As previously argued, handbooks of natural history and science popularization journals during the nineteenth century introduced children and the urban population to racial classification theories, which in turn popularized the idea that the white European was situated at the top of human hierarchies. At the same time, looking at the history of the Hungarian Kingdom, historian Nagy Péter Tibor has rightly observed that the "idea of race" was still in place during the interwar period, due to "the biology teachers who taught morphology and taxonomy along the Linnaeus-Lüben-Gönczy-Pál line" of thinking (Nagy, 2003: 66; Nagy, 2005: 193). Going back to the nineteenth century, a central part of the Lüben-Gönczy pedagogical tradition included the school textbooks authored by the Austrian botanist Alois Pokorny (1826–1886). Pokorny was a teacher who taught biology in the Viennese gymnasium and was the great grandfather of the liberal economist Friedrich Hayek (1899–1992). His biological textbooks, famous for their racialized description of human diversity, were further recommended by Romanian pedagogues working in the schools of Transylvania. When dealing with natural history, Pokorny divided the teaching material into three sections: the first two dealt with the three natural kingdoms, while the third touched upon human anatomy, taxonomy, and morphology. After describing human anatomy in anthropocentric terms, the section dealing with human races explained that "all humans live in society and through sociability he established the moral basis of his family by uniting with its own nation and state." However, differences between the phenotypical features of human races were pointed out in essentialist terms, focusing on factors such as skin color, the form of the skull, teeth, and hair. Coupled with visual representations, the so-called human races were presented in a hierarchical order, starting from Blumenbach's Caucasian race (European, Mongolian, African and Indian), which was followed by several (sub) races. Eventually, Pokorny's school textbook was translated into Romanian and circulated in the Orthodox and Greek-Catholic primary schools of Transylvania (Pokorny, 1910: 189–191).

With regard to teaching natural history in confessional schools, the Greek-Catholic gymnasium at Balázsfalva (now Blaj), was one of the best-organized Romanian secondary schools in Transylvania in terms of teaching facilities. The young botanist and future nature conservationist Alexandru Borza (1887–1971) took the position of natural history teacher in 1911 and reorganized the botanical school garden by introducing one of the first ecological sections called "our forest." In terms of his training, after finishing his theological studies in 1908 in Budapest, he continued in the same city and enrolled in the natural science department at the Faculty of Philosophy (Koszor-Codrea 2023). In parallel, he also conducted a five-months internship at the Secondary School Teacher Training Institute, where the initiator of free education, Pályi Sándor (1859–1929), introduced him to the biological method envisaged by Otto Schmeil. Moreover, according to Borza's student notes, racial classifications were an integral part of the pedagogical exam of the Hungarian educational system. Amongst these included some of the most famous European racial theories of the period, published by Friedrich Blumenbach, Paul Broca, Joseph Deniker, Anders Retzius, and Ernst Haeckel (Fig. 6.2).

The same essentialist notion of race, following Blumenbach's theory, was introduced in the zoology textbooks of the Romanian Kingdom and adopted by famous Moldavian parasitologist Nicolae Leon (1862–1931) by the end of the century (Fig. 6.3). Others went further into the debate by relating racial biological traits to minoritized Romani communities. For instance, pamphlets published by the nature enthusiast Nestor Urechia (1866–1931) addressing children to indulge them in the love of mountains, dehumanized Roma through trivial references to birds such as storks and crows (Urechia, 1924: 49, 64). Racial slurs were not uncommon in the pages of Romanian scientific and folk writings and are still in use up to the present day by the scientific community. For instance, the bird preservationist and natural history teacher, Ion P. Licherdopol (1842–1908), when discussing the popular names

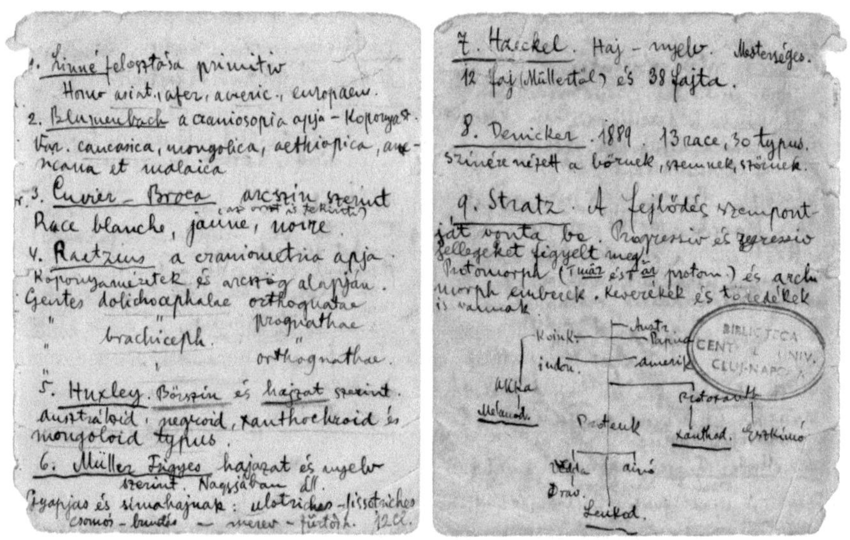

Fig. 6.2 Alexandru Borza's handwritten notes on racial classification theories, created while preparing for the schoolteachers training final exam in Budapest in 1911. *Source* Archive BCU Cluj, Dossier Alexandru Borza, MS6262, Caiet 2

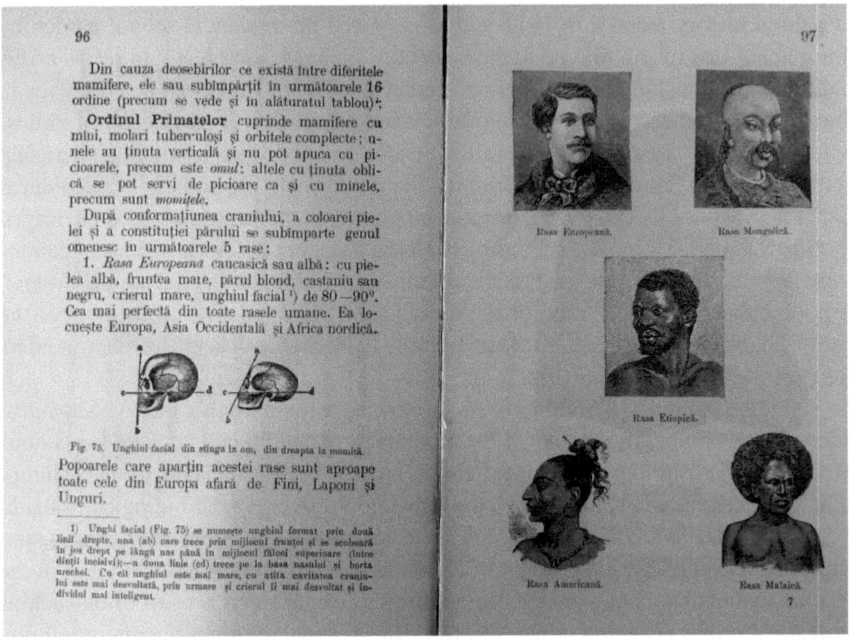

96

Din cauza deosebirilor ce există între diferitele mamifere, ele sau subîmpărțit în următoarele 16 ordine (precum se vede și în alăturatul tablou):

Ordinul Primatelor cuprinde mamifere cu mîni, molari tuberculoși și orbitele complecte; unele au ținuta verticală și nu pot apuca cu picioarele, precum este *omul*: altele cu ținuta oblică se pot servi de picioare ca și cu mînele, precum sunt *momițele*.

După conformațiunea craniului, a coloarei pielei și a constituției părului se subîmparte genul omenesc în următoarele 5 rase:

1. *Rasa Europeană* caucasică sau albă: cu pielea albă, fruntea mare, părul blond, castaniu sau negru, crierul mare, unghiul facial 1) de 80—90°. Cea mai perfectă din toate rasele umane. Ea locuește Europa, Asia Occidentală și Africa nordică.

Fig 75. Unghiul facial din stînga la om, din dreapta la momiță.

Popoarele care aparțin acestei rase sunt aproape toate cele din Europa afară de Fini, Laponi și Unguri.

1) Unghi facial (Fig. 75) se numește unghiul format prin două linii drepte, una (ab) care trece prin mijlocul frunței și se scoboară în jos drept pe lângă nas până în mijlocul fălcei superioare (între dinții incisivi);—a doua linie (cd) trece pe la basa nasului și borta urechei. Cu cît unghiul este mai mare, cu atîta cavitatea craniului este mai desvoltată, prin urmare și crierul îi mai desvoltat și individul mai inteligent.

97

Rasa Europeană. Rasa Mongolică.

Rasa Etiopică.

Rasa Americană. Rasa Malaică.

7

Fig. 6.3 An excerpt from Nicolae Leon's zoological textbook, issued in 1893, depicting Blumenbach's five-race scheme with the "European Race as the most perfect" and supported by anthropometric argumentation. Courtesy of the Mihai Eminescu Central University Library, Iași

of migrating birds such as the glossy ibis, emphasized that its popular name of "țigănuș comes from its various shades of brown, chestnut, dark with bronze highlights, black with lilac, or greenish highlights, which colours its feathers and makes him look like a gypsy by colour" (Licherdopol, 1903: 10).[3] Similarly, when utilizing the introduction of Friedrich Junge's (1832–1905) biological method in secondary schools, Apostol D. Culea (1882–1949) highlighted the racial slurs that were already popular in the Romanian folk culture as follows:

> The crow is the most forsaken and the most mocked bird, for its ugliness, for the sound it makes, and for the harm it brings. All the defects of this bird have been projected on the gypsy, who is called in mockery a crow. When someone is black, it is said he looks like a crow, the same when he sings ugly; he makes the sound of a crow and several other mockeries and hoaxes. To the lăutari, the Romanians are saying 'sing crow from the violin, and from the cobza under the arm' (Culea, 1910: 101).

Finally, race slurs were used as a social and political technique of domination, hence Romanian natural history teachers and ornithologists' views perpetuated the racial discrimination of Romani communities previously put in practice through the institution of enslavement by the Romanian State, its boyars, and the Orthodox

[3] It is important to note that the Ibis was called by Hungarian ornithologists "zsidó-madár" which means the "the Jewish bird."

Church, which existed for a period of five hundred years. Circulating in schools through different textbooks and pamphlets, racial slurs functioned as a vivid collective memory that reproduced the oppressive realities of the enslavement period. When these were coupled with natural history textbooks discussing Blumenbach's racial classification theories, schoolchildren learned negative stereotypes that confirmed the wider white supremacist views about the domination of nature, nation, and race that were prevalent in both Romanian and Hungarian science.

6.5 Conclusions

As shown in this chapter, Blumenbach's legacy in the Romanian region was crucially for shaping and promoting local racial scientific taxonomies. This legacy was made possible by the first generation of physicians such as Iacob Czihac and Iuliu Barasch, who were trained in Germany and ended up working in the Romanian Principalities of Moldavia and Wallachia. Whereas Czihac's text was a direct adaptation from Blumenbach's own work, later manuals were adaptions of secondary German sources, which discussed the same racial taxonomies in simplified versions. Important in many ways, all these narratives overlapped and later reproduced ideas previously part of the oppressive institution of Roma enslavement, still in place throughout the Romanian Principalities of Moldavia and Wallachia up until 1856.

A second generation of native Romanian scholars trained in Paris appeared shortly after the creation of the modern Romanian state, around the 1870s. For scientists such as Ion Ghica and Mihai Obedenaru, racial theories went beyond their mere adaptation to the Romanian language and overlapped with hierarchical, racialized ideas of Romani assimilation and narratives about the Romanians' own origins and "superior" national identity. Similarly, after the institutionalization of physical anthropology in France and Germany between the 1870s and 1914, international scientific networks brought physical anthropologists such as Izydor Kopernicki (1825–1891) and Eugène Pittard (1867–1962) to Romania, who were also eager to contribute to the local debates on scientific racism. Their field research was in line with Romanian and French views on racial anthropology and gave new meaning to the idea of race by focusing on understudied ethnic groups.

Equally important, although the two German and French schools had conflicting scientific ideas in terms of methodological approaches to race, both generations were influenced on the one hand, by Blumenbach's identification of the so-called "Genuine Gypsy" and, on the other hand, they deployed distorted anthropometric measurements and further built on their racialization. Finally, through the teachings of natural history, schools were the main institutions that simplified complex anthropometric racial research for schoolchildren. In doing so, schools legitimized the idea of white anthropocentric domination of both human and non-human species, all the while promoting racial slurs and Blumenbach's racial theories, which uncritically circulate in Romanian biological school manuals to this day.

Acknowledgements I would like to thank the editors of this volume, Victoria Shmidt and Simeone De Angelis, for providing critical feedback and to the anonymous reviewer for important suggestions throughout the various stages of this chapter. I am also grateful to Marius Turda for reading this chapter, to Valeska Bopp-Filimonov for her precious help with translations from German, Iulian Doroftei for transliteration from Cyrillic, Manuela Boatcă for reading recommendations, and to Nicu Adrian Furtună for his patience and further discussions on the history of Roma enslavement in Romania. I would like to acknowledge Constantin Ardeleanu's support of my research endeavors and for his clarifying comments, as well as the librarians at the Central University Library "Lucian Blaga" in Cluj-Napoca for their help. Special thanks are due to the curator and director of the Johann Friedrich Blumenbach—Online Project, Wolfgang Böker, and Dr. Nadine Schäfer, for further providing me with valuable information related to Blumenbach's skull collection and further clarifying part of his correspondence. Any remaining errors are, of course, my own.

References

(1907). Cronica Literară şi ştiinţifică: Fapte. *Convorbiri Literare, 41*(1), 110–111.

Achim, V. (1998). *The Roma in Romanian history*. Central European University Press.

Acton, A. T. (2015). Scientific racism, popular racism and the discourse of the Gypsy Lore Society. *Ethnic and Racial Studies*. https://doi.org/10.1080/01419870.2015.1105988

Baum, B. (2006). *The rise and fall of the Caucasian race: A political history of racial identity*. New York University Press.

Baker, C., Iacob, B. I., Imre, A., & Mark, J. (Eds.). (2024). *Off white: Central and Eastern Europe and the global history of race*. Manchester University Press.

Bărbulescu, C. (2014). Mariage, race et eugenisme en roumanie de la fin du XIX siecle au milieu du XX siecle. *Annales de Demographie Historique, 1*, 235–265.

Berecz, Á. (2013). *The politics of early language teaching: Hungarian in the primary Scools of the late dual monarchy*. Central European University Press.

Bernasconi, R. (2006). Kant and Blumenbach's polyps: A neglected chapter in the history of the concept of race. In S. Eigen & M. Larrimore (Eds.), *The German invention of race* (pp. 73–90). State University of New York Press.

Blakey, M. (1999). Scientific racism and the biological concept of race. *Literature and Psychology, 45*(1–2), 29–43.

Blumenbach, J. F. (1793). Decas altera collectionis suae craniorum diversarum gentium illustrate. *Gottingae, Apud Ioann. Christ. Dieterich.*

Blumenbach, J.F. (1795). *Anthropological Treatises*. Reprinted in 1865 Longman, Green, and Roberts.

Böker, W. (2019). Zur Geschichte der Schädelsammlung Johann Friedrich Blumenbachs. In Deutsche Gesellschaft für Geschichte und Theorie der Biologie (Ed.) *Annals of the History and Philosophy of Biology* Vol. 23 (pp. 3–29) Universitätsverlag Göttingen.

Brosnan, A. M. (2016). Representations of race and racism in the textbooks used in southern black schools during the American civil war and reconstruction era, 1861–1876. *Pedagogica Historica, 52*(6), 718–733.

Caramelea, R. (2015). A school subject serving the nation. Geography textbooks in Romania (1864–1945). *History of Education & Children's Literature, 10*(1), 155–176.

Cheţianu, A. (1902). Istoria naturală şi museul de la şcolele din Blaş. *Tipografia Seminariului.* Arhidiecesan.

Culea, A.D. (1910). *Învăţământul despre natură în şcoala primară*. Tipografia „Neamului Românesc".

Czihac, J. (1837). *Manual de istorie naturală*. Institutul Albinei.

Iordachi, C. (2017). Diplomacy and the making of a geopolitical question: The Romanian–Bulgarian conflict over Dobrudja, 1878–1947. In R. D. Daskalov, A. Marinov & S. Vezenkov (Eds.), *Entangled histories of the Balkans: Concepts, approaches, and (self-) representations* (pp. 291–340). Brill.

Figal, S. (2014). The Caucasian slave race: Beautiful Circassians and the hybrid origins of European identity. In S. Lettow (Ed.), *Reproduction, race and gender in philosophy and the early life science* (pp. 163–168). State University of New York Press.

Furtună, A. N. (2020). From Roma slavery to world war II–Roma resistance in Romania. In A. Mirga-Kruszelnicka & J. Dunajeva (Eds.), *Re-thinking Roma Resistence throughout history: Recounting stories of Strenght and bravery* (pp. 197–215). European Roma Institute for Arts and Culture.

Ghica, I. (1967). *Opere.* Editura pentru literatură.

Ghica, I. (1878). Începutulu Omului. *Annalile Societății Academice Române Memorii și Notițe* Sesiunea II, Tom XI, 61–63.

Gould, S. J. (1996). *The Mismeasure of man.* W.W. Norton & Company.

Grozescu, V. (1885). Din Constanța. *Familia, 21*, 321–322.

Hathaway, J. (2018). *The chief eunuch of the ottoman harem: From African slave to Power-Broker.* Cambridge University Press.

Heathorn, S. (2000). *For home, country, and race: Gender, class, and Englishness in the elementary school, 1880–1914.* University of Toronto Press.

I.M. (1889). Naționalitățile există, măcar de nu ar fi. *Tribuna, IV*(60), 237–238.

Ionescu, M. D. (1904). *Dobrogea in pragul veacului al XX-lea: Geografie matematică, Fisică, Politică, Economică și Militară.* Atelierele Grafice I.V. Socecu.

Istrati, C. (1902). *Activitatea sciințifică a lui Ion Ghica.* Institutul de Arte Grafice Carol Göbl.

Junker, T. (1998). Blumenbach's racial geometry. *Isis, 89*(3), 498–501.

Keel, T. (2018). *Divine variations. How Christian thought became racial science.* Redwood City.

Kolchin, P. (2002). Whiteness studies: The new history of race in America. *Journal of American History, 89*(1), 154–173.

Kopernicki, J. (1872). Ueber den Bau der Zigeunerschädel. Archiv für. *L'Anthropologie, 5*, 267–324.

Koszor-Codrea, C. (2022). Mismeasuring diversity: Popularizing scientific racism in mid-nineteenth century Romania. *Journal for Romanian Studies, 4*(1), 37–56.

Koszor-Codrea, C. (2023). The tyranny of schools: Nature and nation in the schools of Transylvania and the Romanian kingdom, 1870-1914. *New Europe College Yearbook*, 245–280.

Licherdopol, I. P. (1903). *Despre ornitologia Română.* Stabilimentul de Arte Grafice Albert Baer.

McMahon, R. (2009). On the margins of international science and national discourse: National identity narratives in Romanian race anthropology. *European Review of History: Revue Européenne D'histoire, 16*(1), 101–123. https://doi.org/10.1080/13507480802655444

McMahon, R. (2019). *National Races: Transnational power struggles in the sciences and politics of human diversity, 1840–1945.* University of Nebraska Press.

Marino, M. (2022). At the roots of racial classification: Theory and iconography in the work and legacy of Johann Friedrich Blumenbach. *Encyclopédie internationale des histoires de l'anthropologie available online at.* https://www.berose.fr/article2546.html?lang=fr

Mazzolini, R. G. (2014). Skin colour and the origin of physical anthropology (1640-1850). In S. Lettow (Ed.), *Reproduction.* State University of New York Press.

Meruțiu, V. (1906). *Situația școalelor române din Transilvania.* Institutul de Arte Grafice Eminescu.

Morning, A. (2008). Reconstructing race in science and society: Biology textbooks, 1952–2002. *American Journal of Sociology, 114*(S1), 106–137.

Nagy, P. T. (2003). Oktatáspolitikai változások az első világháború után. *Iskolakultúra, 13*(1–7), 66.

Nagy, P. T. (2005). Az állami befolyás növekedése a magyarországi oktatásban 1867-1945. *Iskolakultúra, 15*(6–7), 192.

Pârvulescu, A. and Boatcă, M. (2022). *Creolizing the Modern. Transylvania across Empires.* Cornell University Press.

Petcuţ, P. (2015). *Rromii. Sclavie şi libertatea: Constituirea şi emanciparea unei noi categorii etnice şi sociale la nord de Dunăre, 1370–1914*. Centrul National de Cultura a Romilor.

Piasere, L. (2019). Crania cingarica. La construcción antropológica del cuerpo gitano (1780–1930). *Historia Social, 93*, 103–122.

Pittard, E. (1908). Anthropologie de la Roumanie. L'indice Céphalique Chez 837 Tsiganes (Hommes). Influence de la Taille sur L'Indice Céphalique. In *Asociaţiunea Română pentru înaintarea şi răspândirea ştiinţelor. Congresul şi Expoziţiunea din Bucureşti Anul 1903*. Imprimeria Statului.

Pittard, E. (1913a). Contribution à l'étude anthropologique des Arabes. *Le Globe. Revue genevoise de géographie, 52*, 147–162.

Pittard, E. (1913b). Anthropologie de la Roumanie: les peuples sporadiques de la Dobrodja. VI, quelques disséminés: Tcherkesses, Arabes et nègres. *Bulletin de la Société roumaine des sciences, 22*(4–5), 307–328.

Pokorny, A. (1910). *Istoria naturală pentru şcolile primare*. Editura F.

Puttkamer, J. (2003). *Schulalltag und nationale integration in Ungarn Slowaken, Rumänen und Siebenbürger Sachsen in der Auseinandersetzung mit der Ungarischen Staatsidee 1867–1914*. Oldenbourg.

Raj, B. (2007). The beautiful skull and Blumenbach's errors: The birth of the scientific concept of race. *British Medical Journal, 335*(7633), 1308–1309.

Rambachan, A. (2018). Overcoming the racial hierarchy: The history and medical consequences of 'Caucasian.' *Journal of Racial and Ethnic Health Disparities, 5*(5), 907–912.

Rhode, M. (2019). A matter of place, space, and people Cracow anthropology, 1870–1920. In R. McMahon (Ed.), *National Races: Transnational power struggles in the sciences and politics of human diversity, 1840–1945* (pp. 105–140). University of Nebraska Press.

Roşu, F. (2022). Introduction. In F. Roşu (Ed.), *Slavery in the Black Sea region, c.900–1900: Forms of Unfreedom at the intersection between Christianity and Islam* (pp. 1–18). Brill.

Ruch, M. (1986). *Zur Wissenschaftsgeschichte der deutschsprachigen "Zigeunerforschung" von den Anfängen bis 1900* [Unpublished doctoral dissertation]. University of Freiburg.

Rupke, N. (2019). The origins of scientific racism and Huxley's rule. In N. Rupke & G. Lauer (Eds.), *Johann Friedrich Blumenbach: Race and natural history, 1750–1850* (pp. 232–247). Routledge.

Saini, A. (2019). *Superior: The return of race science*. Bacon Press.

Scheibinger, L. (1990). The anatomy of difference: Race and sex in eighteenth-century science. *Eighteenth-Century Studies, 23*(4), 387–405.

Shmidt, V., & Jaworsky, B. N. (2021). *Historicizing Roma in Central Europe: Between critical whiteness and epistemic injustice*. Routledge.

Treitler, V. B. (2013). *The ethnic project: Transforming racial fiction into ethnic factions*. Stanford University Press.

Turda, M. (2007). From craniology to serology: Racial anthropology in interwar Hungary and Romania. *Journal of the History of the Behavioral Sciences, 43*(4), 361–377.

Turda, M. (2018). *Ştiinţă şi etnicitate: Cercetarea antropologică în România anilor 30*. Editura Muzeul Municipiului Bucureşti.

Turda, M., & Furtună, N. (2021). Roma and the question of ethnic origin in Romania during the holocaust. *Critical Romani Studies, 2*(4), 8–33.

Turda, M. (2022). Războiul Sfânt al Rasei: Eugenia şi protecţia naţiunii în Ungaria 1900-1919. *Editura Academia Română.*

Turda, M. (2024). In căutarea românului perfect: specific naţional, degenerare rasială şi selecţie socială în România modernă. *Editura Polirom.*

Ungureanu, C. (2015). Învăţământul primar şi secundar din Bucovina în perioada 1869–1918. *Akademos, 4*, 118–125.

Urechia, N. (1924). *Fermecătoarea natură*. Tipografia Jockey Club.

Willems, W. (1997). *In search of the true gypsy: From enlightenment to final solution*. Routledge.

William, F., & Dougherty, P. (2012). In N. Klatt, Rev & Augm (Eds.), *The correspondence of Johann Friedrich Blumenbach: Volume IV, 1791–1795* (pp. 645–965).

Cosmin Koszor-Codrea is a postdoctoral researcher at New Europe College in Bucharest and Regional Representative of Romania in the European Society for Environmental History. He received his PhD from Oxford Brookes University in 2021 with a thesis focusing on the transnational dissemination of Darwinism, eugenics and race science in Romania. He also received several grants and awards such as the "Lisa Jardine" award for the history of science from Royal Society of London (2019), nominated as Post-Graduate member of the Royal Historical Society (2019) and co-winner of the "Keith Hitchins" best dissertation award from the Society for Romanian Studies (2022)

Chapter 7
Imperfect Evolution: A Formative Force in Soviet Studies of Oligophrenia

Kseniya Brailovskaya

Abstract This chapter traces ideas of development interlaced with the Russian reception of *nisus formativus* in the eighteenth and nineteenth centuries and their response in the Soviet concepts of oligophrenia in the middle to end of the twentieth century. United by a focus on the transformations caused by damage and disease, as well as the capacity for language and abstraction, descriptions of monsters and studies of oligophrenia are positioned in a specific intellectual tradition of imagining deviation as a means of explaining evolution. I begin by outlining the ideas and contexts influencing the adaptation of *nisus formativus* that were significant for Russian and Soviet understandings of mental abnormality, then focus on the Soviet scientific framing of oligophrenia as a primarily biological phenomenon, supported by its grounding in terms of human evolution.

Keywords Formative drive · Degeneration · Oligophrenia · Mental development · Abstract thinking · Mental retardation · Brain functions · Recapitulation · Psychophysiological pathology · Human evolution · Epidemiological genetics · Defectology · Dysontogeny · Soviet medical genetics

7.1 The Formative Drive: Degeneration, Beauty and Predetermination

The historiography of Friedrich Blumenbach's works on generation situates the formative drive, *nisus formativus*, or *Bildungstrieb,* within the natural-historical quest for the laws of development and critically re-evaluates antiracist attempts to identify human differences. While the accounts of experiments with the regenerating polyp (Blumenbach, 1792) illustrating a common introduction to the eighteenth century discussion on epigenesis, evolution, and the downfall of preformism, their

K. Brailovskaya (✉)
Centre for the History of Science, University of Graz, Graz, Austria
e-mail: kseniia.brailovskaia@uni-graz.at

V. Shmidt and S. De Angelis (eds.), *Blumenbach's Racial Classification*, History, Philosophy and Theory of the Life Sciences 39,
https://doi.org/10.1007/978-3-032-15950-2_7

meaning for human systematisation is less perceptible than studies of skulls and the construction of races.

A focus on generation, degeneration, and regeneration falls within a natural-historical understanding of common qualities amongst all living organisms, whereas the associated anthropological measurements are considering rather the results of a crude reception of theory. However, the image of an ontologically different Caucasian race organising itself in contrast to other races, which emerged through degeneration (Kee, 2019, 138), challenges the very possibility for conceptualization not juxtaposed with a reified sense (Richards, 2000) of developmental differences. Degeneration, even in its primary eighteenth-century meaning as deviation from the original, was an ambiguous tool for assessing the development of posterity as well as the recovery of damaged parts.

Within the history of modern biology, the debates concerning the predetermination of the formative drive outline positions regarding the development, from the idea of perfection to the uncertain nature of transformation. Studies on the mutual influence of Immanuel Kant and Blumenbach in a teleological understanding of generation, along with the history of mechanistic organicism and the separation of biology from physics (Helbig & Nassar, 2016) framing living organisms through a dilemma of design and purpose, resolved in self-regulation (Steigerwald, 2019, 115). Despite Blumenbach's denying the Great chain of being (Grene & Depew, 2004) and the divine plan embedded in living matter, empirically rich experimental descriptions of transformation have shifted the formative drive from the heuristic realm into the Newtonian forces, mysterious but real in their consequences.

The formative force as part of a teleological model without an original ideal pattern or predetermination still had an inner structure. A Kantian understanding of harmony as the unity of the basic laws of nature (Huneman, 2007) as well as of beauty expressed in physiological symmetry and harmony apprehended from experience (Richards, 2019). Along with the focus on organisms in terms of their history, connecting eighteenth century German biology with nineteenth-century evolutionary theory (Sloan, 1979), issues of structure, purpose, and the predestination of development shaped the concept of formative force.

7.1.1 *Nisus Formativus in Eighteenth and Nineteenth Century Russia: Unity and Perfection*

In Russia, Blumenbach likely became widely known after he had won the competition of the Russian Academy of Sciences for works on nutritive power in plants, organised in 1782 in St. Petersburg by Caspar Friedrich Wolff (Dupont, 2007). By that time, Wolff, who long criticised preformism, had developed his own theory of the vital force, *vis essentialis,* associated with nutrition and the capacity for assimilation. In contrast to *nisus formativus,* it was a simple force predisposing growing organisms to future organisation rather than structuring it from within (Dupont, 2007). Although disagreeing on many occasions and differing in their versions of epigenesis, Wolff and Blumenbach both figured monsters to demonstrate alterations (Steigerwald, 2019, 86–87). Wolff's works, however, in the eighteenth and early nineteenth centuries were known to a narrow circle of specialists, unlike Blumenbach's texts published in Russian at the end of the eighteenth century (Blyumenbakh, 1796, 1797).

Russian translations of Wolff, together with a biography, scientific and historical analyses, and, later, in the scientific-popular history (Gangus, 1982), appeared only in the 1950s and 1970s (Gaysinovich, 1961; Vol'f, 1950; Vol'f, 1973), in the wake of interest in the history of evolution theory. First published in 1973, archival documents with the unfinished *Monster Theory* contained the *Fundamentals of the Theory of the Soul* (Vol'f, 1973, 287–288). Wolff described the soul as a material extract of the brain and brain substance of the entire body. It was born and grew with humans and released after death for the true life with God. Although not itself nourished, its growth and perfection depended mainly on the assimilation of substances, during which the simple forces supporting life entered into its forces, from food to the brain, and then to the soul. The quantity and quality of nutrition affected the physical development and susceptibility of the organisms to the monstrosities, variations, and monstrosities-variations that could be transmitted to offspring. But the monster was "an astonishing miracle of God's untold wisdom and omnipotence" (Vol'f, 1973, 52), evidence of the embodiment of an internally or externally undetermined organisation. Bound to a mutable body, the soul remained nevertheless universally human. Wolff's own drawings of conjoined twin monsters and their brains (Fig. 7.1) showed an almost perfect symmetry of form, undoubtedly human, though altered.

> ...Human bodies may alter so ... that ... they will no longer be recognised as human ... However ... what constitutes the true human form will always be preserved in these monsters.... The soul will always remain the same human soul no matter what external form and internal structure the monstrous man receives (Vol'f, 1973, 84-85).

For Wolff, the nourishment of the perfecting soul was an exceptional feature of the emerging, growing, and living animal body. But his reference to Hermann Samuel Reimarus, a honorary member of Saint-Petersburg Academy of Science, and the translators' comments in 1973 that Reimarus conflated children and animals by their lack of memory, as well as absence of capacity for reflection and abstraction (Vol'f, 1973, comments to page 12, 84–85) outline the lasting antique idea of reason as a unique feature of the human mind. In contrasting ability and instinct, Wolff and Blumenbach, who was also attentive to Reimarus (Zammito, 2017), were probably much closer than in explaining the generation. Defying the sensualist concept of cognition, Reimarus opposed human and animal souls on the basis of instincts (*Triebe*), conceptual thinking, and ability to speak. Animals are much more efficient in self-preservation but since they do not operate with concepts, judgments, and inferences, they do not think (Zammito, 2017, 142).

In Blumenbach's popular book on physiology, the lack of most instincts was offset by the capacity for abstract ideas. It developed in the child together with the brain, and determined the other faculties, which, along with speech (Blumenbach, 1795, 115), distinguished man from animals.

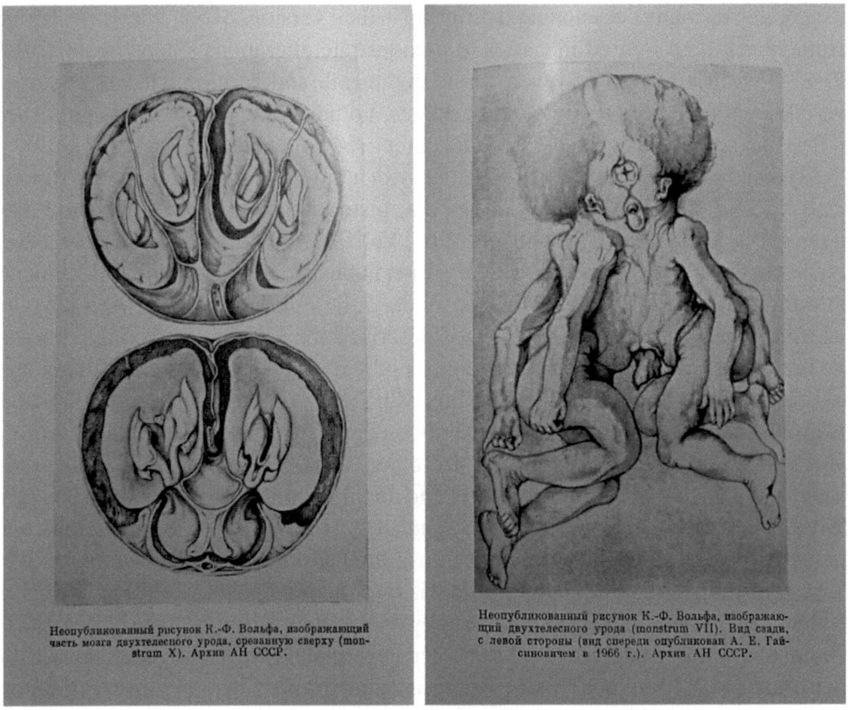

Fig. 7.1 Wolff's drawings from the USSR Academy of Science archive, published with his Theory of Monsters. (*Source* Vol'f, K.F. 1973. *Predmety razmyshleniy v svyazi s teoriyey urodov*. Leningrad. Courtesy of the National Library of Russia, Saint Petersburg)

> After some months, the parts of the brain situated near the anterior-superior region of the forehead, grow more rapidly than the other parts. The forehead, from being flat, becomes prominent, and the child begins to fix its attention upon external objects, to compare, and form abstract ideas, — to generalise (Blumenbach, 1828, 69).
>
> The leading prerogative of the human mind, consists in this, that it alone possesses the exclusive power of reason, by means of which it is able to judge, to form abstract ideas, &c. and which exerts also the greatest influence over most of the other faculties of the mind. In place of this divine power, other animals are endowed with various instincts, or blind and involuntary impulses (Blumenbach, 1795, 199).

Kant, who accepted the idea of the formative drive, went further and defined the ability to abstract as created only by discursive European languages and connected to the particular mindset for grasping moral concepts, fixing the relation between language and human progress (Lu-Adler, 2025).

In the nineteenth century, human development was shaped by an emphasis on health and disease, highlighting deviance in the formative force. Friedrich Haass (Fedor Gaaz) conflated the sense of both Wolff's and Blumenbach's forces in his own understanding of life and health. He was a student of Blumenbach at the University of Göttingen, and stayed in touch after he moved to Russia to practice medicine, where he developed the idea of "assimilative ability," or the capacity of organisms to absorb substances without changing internally and turning them into a part of itself (Mirskiy & Blokhina, 2004). Following his other teacher from Yena University, Friedrich Schelling, in understanding the unity of nature (Wallen, 2004), Haas defined disease as a state of the organism between life and death, both self-preservation and the impossibility to fully assimilate substances. Known in Moscow as "the holy doctor," he was one of the most prominent figures in medicine and philanthropy of the nineteenth century, participating in institutionalisation of public health and creating healthcare inside the penitentiary system, in a sense embodying Schelling's romantic medicine project (Zharov & Chernositova, 1989).

Pavel Goryaninov, a mid-nineteenth century naturalist and medical scientist, made most of the theoretical references to Blumenbach. When he was situated amongst pre-Darwinian evolutionists in the mid-1950s, his variation of *nisus formativus* as *nixus reascendentes* appeared as the projection of upward progressive development (Raykov, 1951, 424–425). Although the Soviet context of the time clearly influenced Goryaninov's image, his understanding of normal and abnormal development indeed combined the formative force with recapitulation.

Goryaninov introduced *nisus formativus* as the latest development in theories of generation in his 1837 book *Zoology: based on zoonomy and applied to general use, especially to medicine, with the addition of a general Organicology and a brief Anthropology* (Goryaninov, 1837), following both Wolff and Blumenbach, making the effect of the formative force dependent on nourishment (Goryaninov, 1837, 18).

> When the formative drive is disturbed, according to Blumenbach, it deviates from the natural order in the development of the parts. And as a result of aberrations of the formative drive (*aberrationes nisus formativi*), monsters, ugly hermaphrodites, bastards, and essential or accidental modifications appear. Human monstrosities are natural, Goryaninov added, because they occur if development is suspended on one of the lower forms changing on the way to the perfect form, and they correspond to the normal shapes of the lower animals (Goryaninov, 1837, 31).

His own version inscribed the process of development into the ascending types of absorption - perception, assimilation, nutrition - as well as into cycles of destruction and regeneration on the way to perfection.

> The original particle of an organic body is gifted with a special power to take nutrients into itself, to resemble its individual organisation, and to reach a certain degree of perfection by continuous destruction, separation and regeneration. Observing animal activity in more perfect and complex organisms, we find that some of them are common to all organic bodies, while others reach perfection only in animals and man (Goryaninov, 1837, 27).

Nutrition was characteristic of the most perfect complex organisms, being the basis of proper development or deviations generating ugly variations and diseases.

> The transformation of the organism that tends to perfection is called progressive (*metam. progrediens s. normalis*); but if it deviates from its path due to natural or accidental, more or less harmful causes, and shows damage or excess, unusual forms and phenomena in organisation in relation to the same or similar bodies, then such transformation is called wrong (*met. anomala*), and its different types: *metamorph retrograda, aberratio nisus formativi,*

> *abortus organorum* etc. This is the basis for the origin of ugliness (*monstrositas*) of freaks or differences (*varietates*), bastards (*hybriditas*) and diseases (Goryaninov, 1837, 29).

Nisus formativus as well as human and animal differences were reflected in the understanding of mind, the germ of which "develops by prudent education to a possible perfection, or is stifled in the savage" (Goryaninov, 1837, 746). Although the mind is inherent in all humans, slavery and lack of education did not allow their mental faculties to advance. For all his abolitionism, however, Goryaninov believed that the full elaboration was possible when farmers and artisans provided for the bodily needs of citizens devoted to intellectual pursuits: mental development had the scale of society, not limited by the individual, and with an indeterminate beginning. The idea of a formative force included animal, human, and overall social development and reverberated his analogy of harmonious unity of nature in the human body.

Complementarity and coherence resonated with Goryaninov's biological systematics developed on the basis of inspired by Schelling Lorenz Oken's idea of intimate biological and historical interrelation of all living organisms (Gambarotto, 2017) and its reimagination in Carl Gustav Carus's classification.

> … Sea sponges can be likened to skin or jointed hair, algae to the ligaments of the uterus or bracteae of plants, fungi to the parts of a flower or the reproductive system, …, i.e. one organ combines the elements of various organs of the most complex organisms…. There are no absolute opposites in Nature - and all the diversity of organisms merges as if into one common organism, harmoniously concentrated in man (Goryaninov, 1837, 107-108).

Yet, man's mind, which "guides him from perfection to perfection" (Goryaninov, 1837, 83), elevating him above the wordless animals, brought a cultural perspective into the organic hierarchy, characteristic of the time. Connections of language, will, and the organic development of the mind among Romantic thinkers of the late eighteenth and early nineteenth centuries were outlined in the theories preceding biological psychology (Richardson, 2001, 5).

Carus, who with his biological systematisation and understanding of the soul, together with Oken (Fig. 7.2), set the stage for an evolutionary version of recapitulation theory, as well as for psychoanalysis (Bell, 2010), saw man's perfection in both the skull and a culturally developed mind (Richards, 2020). However, in contrast to the Carus' division into day, night, and twilight peoples according to their culture and the organic grounding of abilities (Karus, 1844), Goryainov only believed that long exposure to savagery stifled mental abilities, without denying the original mind and the possibility of European enlightenment to all five Blumenbachian races.

The interplay of perfection and degeneration went far from Blumenbach's neutral formative force and more so from Wolff's simple forces which supported the development of body and soul. As both the cause of disease and the drive from one organic form to another, nineteenth century interpretations of *nisus formativus* took on the meaning of a natural and cultural force of development.

Widely known by biologists and anthropologists, Blumenbach's works crossed the boundaries of specialised knowledge. Perhaps most of the future Russian intellectual elite of the nineteenth and twentieth centuries was familiar with Blumenbach's shortcut views through the popular children's textbook, *Natural History*, by

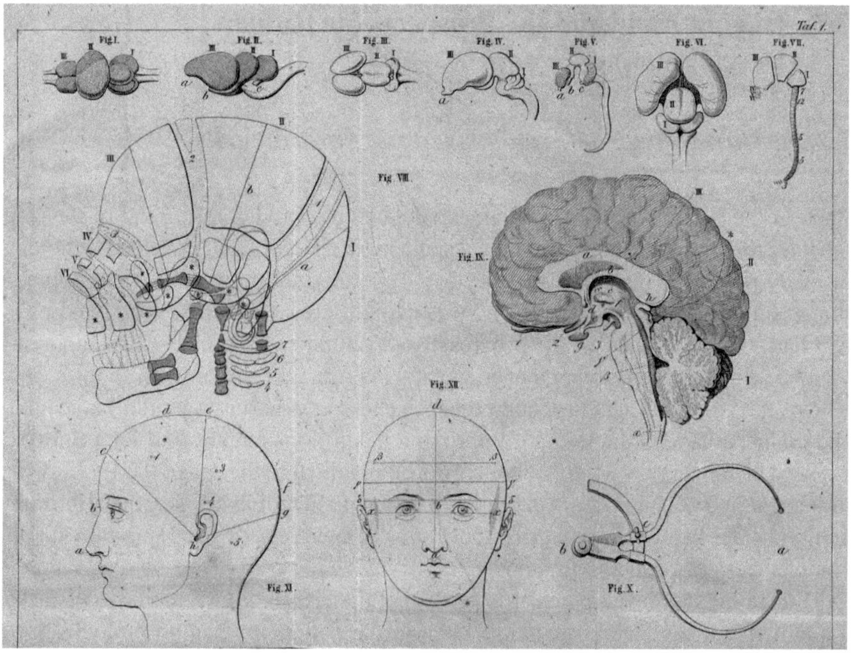

Fig. 7.2 Illustration of Goethe and Ocken's theory, elaborated by Carus, about the origin of the brain from a vertebra in the Russian-language edition of Carus's "Basis of cranioscopy." At the top are images of a brain of a carp, a leguan, a sheep embryo, and human embryos, in the middle are an adult skull and brain, and at the bottom are proportions of an adult human head. (*Source* Karus, K.G. 1844. *Osnovaniya kranioskopii.* Sankt-Peterburg. Courtesy of the National Library of Russia, Saint Petersburg)

Georg Christian Raff, a graduate of the University of Göttingen. He emphasised in the preface the novelty of Blumenbach's approach (Raff, 1778), and the book was published in Russia (Raff, 1796) at the same time as his first translation in 1796, *Physiology or the Science of Human Nature* (Blyumenbakh, 1796). Raff's work, with its pedagogically expounded Linnaeus-Blumenbach system, later became the basis for the natural history part of Konstantin Ushinsky's *The Children's World* (Bezrogov & Tendryakova, 2017), a classic textbook for younger schoolchildren, first published in 1861, the year serfdom was abolished, and republished in the Soviet Union (Ushinskiy, 1948).

7.2 Oligophrenia and the Prospects for Human Development

7.2.1 *Harmony, Comprehensiveness, and Integrity*

Back in the Russian Empire, the nineteenth century moderantist ideal of development, reconciling spiritualism and sensualism, body and soul, called for "the duty of every human being to strive for the full, perfect and harmonious development of capacities" (Kirilov, 1845, 194). Post-revolutionary materialist social and physical harmony embodied in the progressive development and elaborated functions, divided into higher and lower, but connected in norm and pathologies. In the second half of the twentieth century, comprehensiveness cast harmony aside. Its sociality and connection with the tasks of improving institutions emphasised the priority of external factors of psychophysiological development until the end of the 1960s, when biological and social accounts aligned in conceptualizations of integrity. Harmony, comprehensiveness, and integrity took turns as markers of the aspiration for the completeness of form and function.

Human mental development was always at the centre of evolutionary debates as a qualitative difference in mankind, even in the period of "ban on biologization." Ivan Pavlov's suggestion, adopted in the 1930s and reinforced in the 1950s, to study the foundations hidden in the normal state through pathologies (Lubovskiy, 1955), differed from the Marxian view that in the 1920s inspired defectologists to examine the anomalies in child development. Karl Marx's well-known quote that human anatomy is the key to the anatomy of the ape, and the lower can only be understood if the higher is already known, still made sense for the projection of development, in which the oligophrenic state is an underdevelopment of mental functions relative to their adult norm. But pathology, as an opportunity for the manifestation of a hidden basis, offered the comprehension of the norm. The accounts of norm and pathology within the history of the psychophysiological evolutionary development were occasionally focusing on the asynchronous development of organs and functions.

The stability of evolutionarily ancient and newer functions in Soviet experimental psychophysiology became a point of divergence in positions concerning the possibility to compensate for developmental defects. It also resonated with the long-term debate about voluntary action (Jeannerod, 2006), which, in the 1950s and 1960s, was at the centre of theoretical struggles amongst experts, forced to speak in "Pavlovian language" and deal with Pavlov's idea of the inner and outer environment balance (Yaroshevskiy, 1995). At their intersection, an understanding of self-regulation was formed as a human quality rooted in the evolution of the human brain and language. Genetic anomalies later build up on this theorizing in the biogenic chronology of individual (ontogeny) and evolutionary (phylogeny) development. Studies of oligophrenia brought together physiological development, its aberrations, and an evolutionary perspective on human ability. The qualities of evolutionarily ancient and new functions have structured the imagination of oligophrenic development throughout the twentieth century.

7.2.2 Asynchrony of Development

In the first third of the twentieth century, one of the founders of Soviet evolutionary morphology, Aleksey Severtsov, described evolution as changes in organs at the embryonic stage, adding to the pages of phylogenetic history in the works of Jean Baptiste Lamarck and Ernst Haeckel (Severtsov, 1939, 459). Sceptical about the role of heterochrony, a concept defining biological time and evolutionary drivers (Nicoglou, 2017), in the generation of new traits, he recognised its importance for development.

According to Severtsov, in the first stage of morphogenesis, organs are formed, and in the second stage, they develop harmoniously. The closer to the beginning of embryogenesis, the stronger their progressive or regressive phylogenetic changes, leading to an increase and adaptation or to a reduction and disappearance of organs (Severtsov, 1939, 483). The time and sequence of development could also change, reflecting in the correlation for associated growing body parts. Severtsov's student Ivan Shmalhausen later emphasised that the morphology of the embryo is more than just recapitulation, since it reproduces the correlative connections of its actual parts and those that have gone into the evolutionary past, but remained "organisers" of development, such as the chord anlage in vertebrates (Shmal'gauzen, 1982, 71).

Versatile development, conceptualised as the possibility of progressive and regressive deviations at one stage and obligatory harmony at the next, has in a sense fused norm and pathology. The discussion about ancient and new organs introduced into the study of psychophysiological functions assumption about their different evolutionary ages. Associated developmental epistemologies have often implied that pathologies lead not merely to deviance, but to evolutionary recoil. However, in the 1920s, it did not have a strictly negative meaning and rather created a point of equilibrium in assessments of viability.

For more practical discussions, human physiology raised a biogenetic question about the evolutionary chronology of organs and systems and its importance for proper functioning. In 1927, Soviet physician and theorist of "evolutionary pathology," Moisey Lorin-Epstein, gave the most elaborated answer, calling it the basic aetiological principle, according to which "the stability of organs, their parts, characteristics, etc.... and their ability to recover, ... are proportional to their mutual phylogenetic ancientry" (Belkaniya et al., 2017, 158). This definition was related to his understanding of races and their hybrids, which, due to variability of the endocrine system, had different capacities to complete the metamorphosis of phylogenetic shapes to the latest versions. Basically denying the Mendelian rule for the transmission of traits, Lorin-Epstein made the endocrine system responsible for ontogenic speed affecting the organs on their development from past to future shapes.

Like Goryaninov, he saw the basis of most human changes, although rather negative for health, in upright posture. Unstable phylogenetically young organs, compared to ancient ones, were more fragile and vulnerable to pathologies because of insufficient fitness to bipedalism. Lorin-Epstein referenced Robert Wiedersheim, who listed 86 rudimentary organs, and Paul Mathes, who established the forms of the past and future in the female constitution (Fig. 7.3), aligned with a reproductive system of women highly influenced by the environment. Mathes stated that the ultimate principle of human variability is the weak or more often strong ability of buds or chromosomes to develop, creating complexes of traits belonging to the past or the future (Mates, 1927). The form of the past is more realised in female bodies, and the form of the future was reflected in the male constitutions. The criteria for belonging to the future were upright posture, emancipated arms, and predominance of the brain over the entire nervous system. Perfecting Mathes'vision, Lorin-Epstein identified theoretical types that shaped complex variants in a spectrum of normal and pathological development.

> The type of the future is complete recapitulation and complete inheritance of progressive characteristics … is the least stable and often exhibits a number of pathological conditions…. In the natural order of things, these individuals are subject to weeding-out (elimination) in the process of natural selection; culture preserves them as pathological variations…. The type of the past represents a somewhat incomplete recapitulation when inheriting a large number of characteristics common to our direct ancestors. This form is harmonious, capable of reproduction and, like the more ancient type, more stable (Belkaniya and Shpukal 2017, 158).

The aetiological principle was not limited to physiology and applied to reflexes as well. Decades later, Lorin-Epstein's views were still relevant, although "deracialized" and focused on the human ability to change the environment and thus influence the functioning of the body (Ginzburg, 1963, 55–58). Together with the physiological understanding of harmony as a necessary quality for certain stages of development, the need for functional balance and its explicable timeline within the individual body outlined frames of reference for anomalies.

7.2.3 *Oligofrenia in the Balance of Ancient and New*

The Soviet understanding of oligophrenia as an imbalance of evolutionary old and modern functions took shape after the Second World War, but its ideological, theoretical and empirical foundations appeared much earlier. The idea of recapitulation, emerging since the late nineteenth century as connected with evolutionary theory, inscribed the stages of development in a history of transformation of shapes and abilities from the lowest to the highest (Brailovskaya, n.d.). Therefore, the concept of "dysontogeny" ("dysontogenesis"), associated by Soviet experts with the abnormalities of embryonic development described by Gustav Schwalbe, appeared in the evolutionary perspective of the specialised literature since the 1950s. Along with identifying the main syndrome of oligophrenia as an impaired capacity for

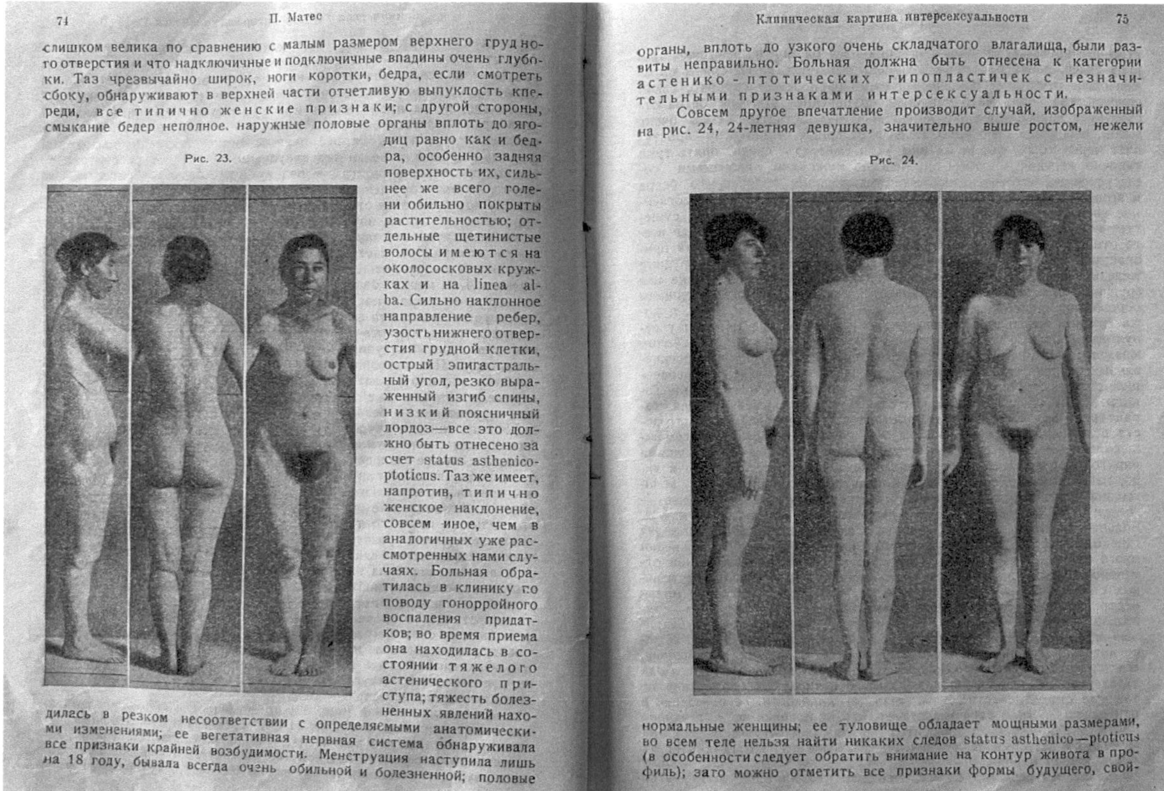
74 П. Матес

слишком велика по сравнению с малым размером верхнего грудного отверстия и что надключичные и подключичные впадины очень глубоки. Таз чрезвычайно широк, ноги коротки, бедра, если смотреть сбоку, обнаруживают в верхней части отчетливую выпуклость кпереди, все типично женские признаки; с другой стороны, смыкание бедер неполное. наружные половые органы вплоть до ягодиц равно как и бедра, особенно задняя поверхность их, сильнее же всего голени обильно покрыты растительностью; отдельные щетинистые волосы имеются на околососковых кружках и на linea alba. Сильно наклонное направление ребер, узость нижнего отверстия грудной клетки, острый эпигастральный угол, резко выраженный изгиб спины, низкий поясничный лордоз—все это должно быть отнесено за счет status asthenico-ptoticus. Таз же имеет, напротив, типично женское наклонение, совсем иное, чем в аналогичных уже рассмотренных нами случаях. Больная обратилась в клинику по поводу гонорройного воспаления придатков; во время приема она находилась в состоянии тяжелого астенического приступа; тяжесть болезненных явлений находилась в резком несоответствии с определяемыми анатомическими изменениями; ее вегетативная нервная система обнаруживала все признаки крайней возбудимости. Менструация наступила лишь на 18 году, бывала всегда очень обильной и болезненной; половые

Рис. 23.

Клиническая картина интерсексуальности 75

органы, вплоть до узкого очень складчатого влагалища, были развиты неправильно. Больная должна быть отнесена к категории астенико-птотических гипопластичек с незначительными признаками интерсексуальности.

Совсем другое впечатление производит случай, изображенный на рис. 24, 24-летняя девушка, значительно выше ростом, нежели нормальные женщины; ее туловище обладает мощными размерами, во всем теле нельзя найти никаких следов status asthenico—ptoticus (в особенности следует обратить внимание на контур живота в профиль); зато можно отметить все признаки формы будущего, свой-

Рис. 24.

Fig. 7.3 Illustrations from the Russian translation of Paul Mathes'book, showing the transitivity of intersexual unstable forms of the past (left) and more stable forms of the future (right). (*Source* Mates, P. 1927. *Konstitutsional'nyye tipy zhenshchin, v chastnosti interseksual'nyy tip*. Kiyev. Courtesy of the Russian State Library, Moscow)

abstract thinking, disontogeny began to refer oligophrenic development to its physiological and social evolutionary incompleteness, in many ways continuing the pre-revolutionary studies.

Abstract thinking in terms of the fitness of academic disciplines for mental development, has been a traditional argument in debates on education since the eighteenth century. In the nineteenth century, it was reinvented in psychopathological descriptions such as Eduard Seguin's, whose work illustrated to Soviet experts the link between the absent abstraction of idiots and the lack of will (Sukhareva, 1962). This image was also empirically supported by Ivan Mierjeiewski, one of the founders of Russian psychiatry known for his studies of idiocy. In 1881–1882, he supplemented the three-volume textbook of psychiatry by Richard von Krafft-Ebbing, written on extensive empirical material, with his own observations (Kraft-Ebing, 1882). The distinction between idiocy and imbecility in the ability to form concepts combined Austrian and Russian data into a well-grounded theory. For Soviet experts, Mierzejewski remained a significant figure also because of the dilemma of an ontogenic and phylogenetic focus in understanding the origins of oligophrenia, positioning microcephaly as proximity to apes for Karl Focht, or to the human embryo for Mierzejewski (Pravdina-Vinarskaya 1957). Oligophrenia as an incomplete biological and social development in the Soviet historiography connected the pre- and post-revolutionary period in the recapitulation model.

When Grigory Troshin published the two-volume *Anthropological Bases of Education. Comparative Psychology of Normal and Abnormal Children* in 1915, dedicating it to Konstantin Ushinsky, he believed that a biological perspective was most suitable for special psychology. He regarded judgement, or the act of ordering reality, as an adaptation to the environment, considering the thought of idiots, imbeciles and debiles as a part of evolution and associating underdevelopment with its previous stages (Troshin, 1915, 69–71). Books aimed at a complete picture of variations in psychological development included all the classic chapters, such as sensation, perception, the ability to generalise, speech, perception of time and space, will, memory, aesthetic and social feelings, etc. The rich description of children's norms and deviations was focused on the evolution from primitive-primal to higher processes. Troshin's work remained significant for Soviet studies of oligophrenia, which, however, were often less integrated into general psychology and focused on more detailed descriptions of physiological features and the aetiology of deviance.

Although studies of mental retardation were carried out by a variety of specialists in the post-revolutionary period, texts devoted specifically to oligophrenia began to appear mostly in the mid-twentieth century. This was largely facilitated by policies aimed at controlling homeless children and integrating the retarded and disabled into education and work. The new challenges required engaging all possibilities of development, reproducing classical ideals and drawing on the latest scientific advances. Oligophrenia emphasised the unity of body and mind, revealing how unsuccessful physiological development affects the shaping of the central nervous system and cognitive abilities.

In Philadelph Zabugin's book *Oligophrenic Children (Imbeciles and Idiots)* (Zabugin, 1937), published in 1937, at the peak of the Great Terror and after the ban on biologically determined explanations of human development, harmony appeared as a successful combination of all elements of personality. Its physiology was explained through the localization of brain functions and visualised, as Zabugin inscribed, in the scheme of Emil Abderhalden (Fig. 7.4), the foreign member of Academy of Sciences of the USSR and president of German Academy of Sciences Leopoldina, associated with Soviet scientists (Kaasch & Kaasch, 2003). The scheme linked brain centres, sensorimotor functions and the abilities to write and speak, showing a properly organised nervous system. Harmony as a goal of psychophysiological development combined the adjustment of the human brain with development of the body and feelings.

> So, on the one hand, the intimate connection of the autonomic nervous system with organs (lungs, heart, kidneys, liver, spleen, etc.), on the other hand, the controlling work of consciousness (CNS) in cases of complete consistency (sodruzhestvennost') creates harmony of the individual; in cases of the slightest violation of such consistency, we get a ladder of types of children who deviate from the average norm. Life experience and personal harmony are the result of the growth of all parts of the nervous system, but mainly are a result of the growth of consciousness and subcortical centres with which the autonomic nervous system is connected…. in order for this harmonious activity of intelligence and emotions to occur, it is necessary to have certain areas of the brain in a mature and healthy state (Zabugin, 1937, 42-43).

Zabugin's attempt to consolidate knowledge about oligophrenia for educational purposes from the perspective of scientific criteria was hardly regarded as serious. However, its focuses and theoretical visualisation were quite in line with the objectives of oligophrenia research in the following decades, which reinforced an evolutionary perspective connecting biological and social angles.

Distancing from the bourgeois and too biologically oriented pedology and defectology, virtually prohibited after 1936, researchers manoeuvred between the ban on biological and social determinism and the need to create the necessary tools for child development. Since anomaly articulated in the practical tasks of more accurately assigning children to institutions, physical evidence of deviations had become useful. In his 1939 book *Psychology of the Mentally Retarded Child*, Leonid Zankov emphasised deviations in the structure from the very beginning of the abnormally developing brain of oligophrenics, with pathologically simple, too intermittent, or merging cerebral fissures, with fewer small nerve cells, and in general more typical for embryos:

> The material basis of the deep uniqueness of the mentally retarded child's psyche is anomalies of the structure and activity of the central nervous and endocrine system. Studies of the central nervous system of deceased oligophrenics have revealed insufficiency of the brain of oligophrenics from various sides. Thus, pathological deviations in the structure of the furrows and gyrus of large hemisphere cortex were found. The pathological simplicity of the furrows and gyrus is noticed…. The furrows of the large hemispheric cortex of severe oligophrenics often resemble a pattern characteristic of a certain stage of embryonic development (Zankov, 1939, 10-11).

The association of an anomaly with an ontogenical timeline as empirically based argument defined the types and prognosed their future. An evolutionary perspective

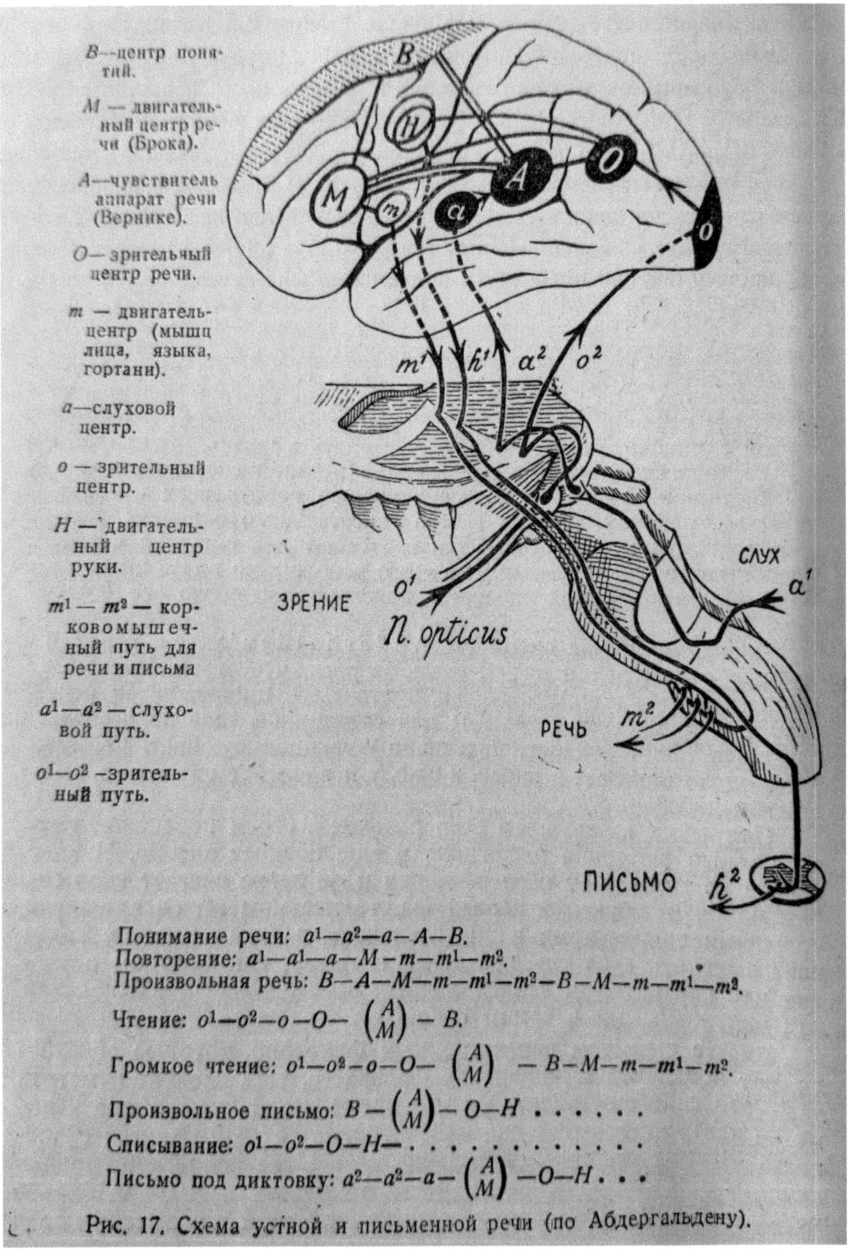

Fig. 7.4 Scheme of oral and written speech (according to Abderhalden). (*Source* Zabugin, F.D. 1937. *Deti-oligofreny (imbetsily i idioty)*. Moskva. Courtesy of the Russian State Library, Moscow)

supported the idea of oligophrenia as a complex of related developmental deficiencies, which was however not fully realised until the second half of the twentieth century.

In the 1950s, interest in oligophrenia resurfaced, finding a basis for psychophysiological and later for genetic theories. Elena Pravdina-Vinarskaia in her 1957 *Neurological characteristic of the oligophrenia syndrome* (Pravdina-Vinarskaya 1957), described oligophrenia as a coherent defect that impacts vision and hearing, motor skills, and sensitivity and concluded that it is necessary to study not individual elementary functions, but entire functional systems. Two years later, Maria Pevzner argued that instead of studying numerous symptoms, it is useful to focus on the main symptom, to establish a pattern, and to look at the structure of the defect as a whole, creating true natural science correctional education.

Pevzner in her 1959 book *Oligophrenic Children (Study of Oligophrenic Children in the Process of Their Upbringing and Education)* (Pevzner, 1959), published in English two years after (Pevzner, 1961), outlined a category of children with "severe deficiency" of the frontal cortex. It is here, she believed, that most researchers located phylogenetically newer higher human functions associated with "underlying" functions. Like many others (e.g. Luriya, 1958), Pevzner defined it as underdevelopment of the extracting and generalising function, that is, analysis and synthesis, or thinking in concepts. Pathology, mainly occurring after diseases or injuries in the embryonic period or in the first months of life, halted the ontogenetic development that should have repeated the psychophysiological evolution of the frontal lobes, increasing in size and connecting to all parts of the brain.

> Comparative anatomical data show that the frontal lobes belong to the most recent formations of the brain. There is a significant increase in the frontal lobe in the ascending phylogenetic track, especially in humans. The frontal cortex is connected by associative ties with all areas of the brain. Fibers from the frontal cortex stretch not only to the cortex of other areas of the hemispheres, but also to the underlying formations. The richness of ties in the frontal part suggests its particularly great functional significance. The frontal lobes emerged last in a long process of evolution. In humans, they occupy up to a third of the cerebrum. Thus, it becomes clear why higher specifically human forms of behavior were associated with the function of the frontal lobes (Pevzner, 1959, 376).

Pevzner opposed the psychological "preformism" of the inherent embryonic qualities to exogenous developmental factors, considering an approach on the basis of the latter more progressive, since it included the part of human evolution beyond biological development. Referring to the work of the Moscow Brain Institute, she described as anomaly the inability to operate with concepts and the physiological underdevelopment of the brain - the impossibility in ontogenesis to complete the phylogenetic construction of all layers and functions of the brain, from the ancient to the newest.

When in 1958, in a collection of experiments with mentally retarded children conducted under his supervision, Alexander Luria demonstrated the power of the evolutionarily younger function of speech, examples of oligophrenia set its boundaries (Luriya, 1958, 38). He also stressed the importance of studying oligophrenia as a peculiar condition to narrow the clinically diverse category of mental retardation (Luriya, 1963, 22).

For Luria, speech created a system where "the neurodynamically most perfect and formed element acquires the role of a controlling mechanism" (Luriya, 1958, 21–22). And it was not just the most phylogenetically new mechanism, but better and more stable than the previous ones, excluding cases of severe oligophrenia.

> … The word is always a complex stimulus, including both non-specific (inducing or inhibiting) and specific (selective, meaningful components) stimuli; we have seen how in evolution the former gradually recede into the background and give way to the leading, dominant place to the latter. The specific feature of that functional systems defect which characterises oligophrenia is precisely in the fact that this predominance of selective (abstracting and generalising) word connections is formed here too insufficiently and primitive non-specific word functions easily begin to dominate (Luriya, 1958, 39).

Oligophrenics embodied severe damage to the upper layers of the cerebral cortex at the very beginning of brain development, resulting in incomplete thinking. Their mind was chained to the senses and the defective neurodynamic, disbalancing speech and motor connections. Speech abilities have specified the perception of oligophrenia and the boundaries of inevitably abnormal development:

> Can we then expect that an attempt to include this defective in its structure and inert in its dynamics speech as a regulating mechanism can lead here to any significant effect? … Could we expect that the permanently inhibitory cortex of severe oligophrenics would allow us to turn to their defective speech as a compensatory means? (Luriya, 1958, 38).

Pevzner made similar division, believing that many limited, stupid, and incapable children (Pevzner, 1959, 5), or children who are retarded because of social causes (Pevzner, 1966) are inappropriately categorised as oligophrenic. Further work by academic scientists like Pravdina-Vinarskaya, Pevzner, or Grunya Sukhareva, one of the founders of expert knowledge on oligophrenia, organised theory and a quickly growing research practice. The opposition of evolutionary new, and therefore fragile to old and therefore stable, functions structured the accounts of oligophrenic development.

As Sukhareva noted in her lectures for advanced training courses of child psychiatrists, in idiocy, the deepest degree of mental retardation, "all mental life is in its embryonic state" (Sukhareva, 1961). Diffuse and total oligophrenic feeblemindedness affects all processes of cognitive activity and is combined with the underdevelopment of personality, motor deficiency, and often with the underdevelopment of organs in general. At the same time, "[M]ainly ontogenetically and phylogenetically young functions are affected, whereas more ancient functions remain relatively intact. This pattern is general in the pathology of the central nervous system" (Sukhareva, 1961, 11). Oligophrenic development also impaired the accuracy and gracefulness of movements, made impossible elaboration of some senses and complex emotions as well as associated volitional purposeful action, and in cases of severe idiocy led to "elogia" - the inability to think. The focus was on the development of the central nervous system and any aetiology of oligophrenia dovetailed into a pattern of old and modern functions.

In textbooks and lectures from 1961–1962, published three years later in one volume (Sukhareva, 1965), Sukhareva included the basics of genetics to introduce scientific descriptions of hereditary causes. Despite the different origins of developmental abnormalities, all variations of oligophrenia added up to a common picture of disontogeny of the nervous system, or disrupted ontogenesis. There is no sharp boundary between exogenous and endogenous factors, she argued, based on studies of various pathological influences, from viruses and toxic substances to the brain underdevelopment caused by radiation in Hiroshima. Because in all cases, there is a metabolic disturbance with a time difference - the germ cells of parents or ancestors are affected. Therefore, birth dysplasias, i.e. obvious physiological abnormalities, are similar to hereditary ones, which could be explained by the concept of phenocopies (Sukhareva, 1962, 5–6).

7.2.4 *Genetic Turn in Studies of Oligophrenia*

In the 1960s and 1970s, Soviet research on oligophrenia diversified in its methods and its combination of genetic, psychophysiological, and psychiatric perspectives, mainly focusing on detectable physiological abnormalities (Fig. 7.5). The World Health Organization's interest in human genetics and intellectual disabilities legitimised mass screening in special institutions, schools, and specific locations for prevention and early intervention. Diverse data and interpretations, enhanced by access to international sources, required updates in the understanding of oligophrenia. But this was complicated by the fact that the majority of cases belonged to the so-called undifferentiated oligophrenia, in which the cause could not be determined, and even the differentiated cases often implied assumptions rather than a well-defined aetiology.

The theories of phenocopies and genocopies, combined with the idea that developmental problems depend on the time periods of development and harmful exposure (Pevzner, 1972), balanced the inability to establish an aetiology in most cases of oligophrenia. Such conceptualisations, however, were in line with an understanding of oligophrenia as a disorder of a consistent process that, given all the right conditions, tended towards a predictable outcome, whether normal or abnormal. Its holistic sense adjusted later polyaethiological and heterogeneous descriptions, supported by explanations of familial accumulation of numerous different possible causes of oligophrenia or a combination of inherited disorders and adverse environments (Yavkin, 1974). The understanding of oligophrenia as lagging behind the evolutionary modern state of psychophysiological functions remained unchanged despite the emergence of new classifications unifying aetiology and pathogenesis.

Thus, in the 1975 collection published by the Research Institute of Psychiatry of the Ministry of Health, the classification of oligophrenia had types built in the evolution of cognitive activity according to the criterion of damage to evolutionarily new and old systems (Portnov, 1975). In a way, the book became an intermediate result of many years of biochemical screening, and psychophysiological, pedagogical, cytogenetic, genealogical, psychiatric, neurological, and various other studies,

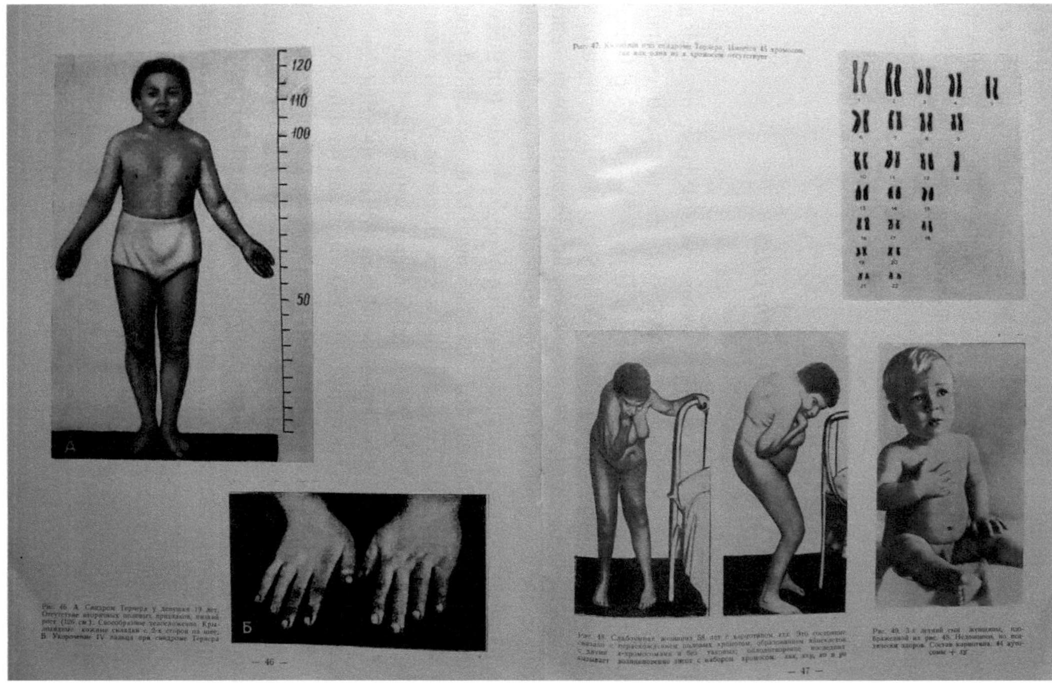

Fig. 7.5 Set of pictures from the training guide for a psychiatry course on oligophrenia and its border cases with other diseases. In the photos are a girl, with her hands and chromosomes as the illustration of Turner syndrome, and a mentally retarded woman with chromosome damage with her normal son. (*Source* Russkih, V.V. 1969. *Oligofrenii i smezhnyye formy*. Moskva. Courtesy of the Russian State Library, Moscow)

through which tens of thousands of children diagnosed with oligophrenia and their families had passed. The understanding of the oligophrenic defect as only in a small number of cases monogenically inherited, and in many cases polygenic and multifactorial, extended it beyond the limits of individual development. The damage affected an individual organism, but the aetiology, shaped by an epidemiological framework, involved the accumulation of the "genetic load" due to frequent inbreeding of isolated groups, facilitating the pooling of deleterious mutations leading to metabolic abnormalities, which could also be aggravated by trauma, disease, or harmful environments. Determining whether, in each individual case, an external influence such as neuroinfection, toxic exposure, maternal Rh factor, radiation, hypoxia or hereditary factors became the leading cause of oligophrenia could become a potential diagnostic possibility in future.

The epidemiological framework in the 1960s and 1970s shaped perceptions of abnormalities underpinned by the parallel development of medical geography. As an international trend and the basis for the joint statistics of medical, psychiatric, and psychological assessment of populations around the world (e.g. Bash & Bash-Liechti, 1987), emerging genetic epidemiology and the works of one of its founders, Newton Morton, became a legitimate reference for Soviet experts. Seeking to encompass all possible sources of oligophrenia, they combined it with family studies such as the Elizabeth and Sheldon Reed project, continuing the research of Charles B. Davenport (Reed & Reed, 1965). And Lionel Penrose's understanding of mental defects being an interplay of heredity and environment encouraged a shift in focus, from the earlier interpretation of oligophrenia as primarily an externally caused disease (Pevzner, 1966, 11).

Isolates, socially and geographically closed populations,[1] were of particular interest to account for all factors leading to oligophrenia as much as possible, and such a study (Barashnev & Kazantseva, 1975) was included in the 1975 collection, as well as in a collection published by the Ministry of Healthcare on clinical genetics and its practical use in paediatrics (Barashnev & Vel'tishev, 1975). A locality in the Komi Republic attracted attention because of the frequency of medical and mental abnormalities, and years of research expeditions began. Paediatricians, psychiatrists, neuropathologists, a psychologist, biochemists, and health organisers made genealogical, clinical, and laboratory studies of children and some of their relatives. Rural genealogies had been investigated since the first half of the eighteenth century to estimate the index and the percentage of inbreeding. A paediatrician, a psychiatrist, a neurologist and a psychologist examined organs, systems, physical development and did anthropometric measurements. The psychologist used "experimental methods, graded by age, to judge the state of thinking, memory, mental efficiency, learning ability, attention (Kraepelin test, Luria method, Bernstein's "sequence of events" method, blank method, learning experiment method, Seguin method, adapted Wechsler method)" (Barashnev & Kazantseva, 1975, 154).

[1] Similar studies were, for example, conducted earlier in Sweden (Sjogren & Larsson, 1970).

Laboratory diagnostics, which included sex chromatin testing, blood groups, qualitative urinalysis, amino acid concentration and quality, and review of existing medical records, was impressively broad.

Additionally, dermatoglyphics[2] was studied, showing that patterns on the finger skin of mentally retarded children and their relatives differed from the control group. Amongst the many physiological abnormalities of both normal and mentally retarded children, such as impaired hearing, vision, skin, teeth, muscles and bones, the researchers particularly highlighted the stigmas of dysembryogenesis - the gothic palate, hypertelorism, anomalies of the ears and so on. The similarity of anomalies in families and the statistics of mental retardation showing that it was six times more frequent in inbred marriages, allowed researchers to assume that there was a hereditary nature to oligophrenia in the village. In 1983, the outcome of expeditions between 1971 and 1978 in three Komi villages reflected the researchers' confidence that oligophrenia of unidentified aetiology was largely the result of hereditary biochemical abnormalities (Barashnev et al., 1983). Between the two descriptions of the northern isolate in 1975 and 1983, oligophrenia studies took a turn toward largely genetically based explanations.

The tasks of developing medical and genetic counselling on mental retardation, set back in the 1960s, were carried out not only in oligophrenia studies, but they also had theoretical support in medical genetics. Vladimir Efroimson has been developing its clinical part since the mid-1960s, after the victory over Lysenkoism. Efroimson's *Genetics of Oligophrenia, Psychosis, Epilepsy* (Efroimson & Blyumina, 1978) written with Marina Blyumina in 1979, united Soviet and international theory and practice. World statistics drew out the main causes of oligophrenia, aligning various possible factors and psychodiagnosis with IQ measurement:

> The foreign literature of recent years has paid much attention to the role of child nutritional disorders (in the first years of life) in the development of mental retardation. It has been found that in South Africa, children who had severe dystrophic disorders in early childhood were mentally retarded compared to their peers despite recovery of their physical condition in the following years. Their IQ (intelligence quotient) by school age was 20 units lower than that of children who had not suffered debilitating conditions in early childhood. Similar results were obtained in Mexico and Guatemala (Efroimson & Blyumina, 1978, 18).

Bad environment, however, rather allowed bad heredity to manifest itself as the impaired brain of a child or embryo. Since the formation, development, and activity of the central nervous system is controlled by a multitude of genes important at different stages of ontogenesis and after its accomplishment, mental normality is ensured only by the correct and coordinated in time action of various loci responsible for the foundation, shaping, development, and functional completeness of the brain (Efroimson & Blyumina, 1978, 3). Taking Sukhareva and Pevzner's understanding of oligophrenia as a basis, Efroimson and Blyumina elaborated a genetic and psychophysiological picture of the step-by-step process of brain development and its pathology leading to the damage of late-forming higher functions:

---

[2] Dermatoglyphics developed in the USSR mainly within anthropology (Khit', 1983).

> An important feature of brain ontogenesis is a large gap in the timing of maturation of its different sections, as well as different basal nuclei and even cortical layers. In the same stage of ontogenesis it can be seen an unequal degree of differentiation between various brain systems and unequal stages of development of nerve elements within the same systems. This fact determines the almost inevitable brain damage of the foetal embryo and infant when exposed to genetic or exogenous harm at any stage of the central nervous system development. The structure of mental defect in oligophrenia, relating to the time of brain damage to different stages of ontogenesis, can differ significantly. But in all cases, the late maturing brain systems responsible for higher mental functions are particularly affected (Efroimson & Blyumina, 1978, 23).

In fact, despite the recognition of damaging environmental factors, the leading role of genetics in the emergence of oligophrenia, the authors asserted, is beyond doubt, since pathological heredity has a crucial role in the at least 80 per cent of cases of moderate and mild mental retardation and 50 per cent of cases of severe and profound general mental underdevelopment. A year later, a book edited by Pevzner (Pevzner & Lebedinskaya, 1979), although attempting to return the focus to exogenous factors in oligophrenia, so richly described the abnormalities, embedded in the gradual development of the brain, that it rather complemented genetically rooted explanations. In 1986, Blyumina claimed that because of the complexity of undifferentiated cases, defectology should use population genetic studies with clinical, clinical genetic, statistical genetic and mathematical analysis (Blyumina, 1986, 4), referring to completely different ways of assessing abnormal development.

The understanding of oligophrenia as a failure of a consistent step-by-step organisational process, combining bad genes, external influences, the disorder of brain structure, and inability to develop abstract thinking in its ontogenesis, went through a period of changing focus from function to structure and back again in late and post-Soviet period, and has generally remained unchanged until now.[3] The harmony of the first half of the twentieth century, however, was mostly lost, along with its humanitarian part.

7.3 Conclusion: Unfinished Evolution

Critical studies emphasise the teleology behind Blumenbach's and his followers' theorising the shapes as goals and the blurring line between epistemology and ontology in the imagination of the generative process via the concepts such as formative drive. The Soviet case of oligophrenia studies also shows that physiological descriptions of deviance were essential in the reception and adaptation of *nisus formativus*.

Illustrating the transformation and visual differences produced by degeneration and regeneration, Blumenbach's idea of the formative drive helped to assemble more

[3] Notably, on the cover of a recent textbook titled *Mental retardation (oligophrenia)*, the authors placed an image of René Magritte's 1926 'Polar Light', depicting torn fragments of naked female figures against a backdrop of sand dunes (Kolmogorova & Buykov, 2013).

complex phenomena into integrity, based on an individual's belonging that has never been limited to physiology and anatomy. As part of human generation, it explained the difference between humans and animals through instinct, abstract thinking and language. In the eighteenth and nineteenth centuries, *nisus formativus* became an element of the recapitulation of naturally and historically aligned biological forms.

Development as the striving for harmony, which includes functional stability, internally rooted but affected by external conditions, disease, and damage, unites the epistemologies of Soviet studies of oligophrenia in the twentieth century and the Russian reception of the formative drive in the eighteenth and nineteenth centuries. As the discussion on evolving functions in the twentieth century combined criteria for shaping and development, the idea of stable ancient and fragile new functions merged the imagination of changing forms with the established intellectual habit of emphasising progressive and exclusively human mental faculties, above all, abstract thinking.

Oligophrenia as unfinished development upheld binding of form and function, which, since the middle of the twentieth century integrated the language of physiology and genetics. But, although emancipated in the second half of the century, rapidly developing Soviet psychology barely dealt with the boundaries of mental development that were difficult to bring into the understanding of human progress and human species in general. The humanistic ideal in the mid-twentieth century, assuming that even in the most severe cases people were able to develop, could not accommodate the variety of new ways to describe anomalies. Besides, it barely fit for purpose as it was underpinned by a vision of evolution that placed humans simultaneously in the cultural, physiological, and genetic dimensions of an epistemology based on the ideal of the completeness and coherence of forms and functions. Such a vision excluded a number of them from a version of human history as the unfolding capacity of the mind to master the abstract constructs.

Defining the force affecting psychophysiology as external, Soviet scientists in general supported the assumption that the difference between humans and other species was due to the fact that the physiology of the human brain is evolutionarily linked to the ability to analyse and synthesise, impossible without language. Biological and social development was embedded in the interpretation of the norm and outlined by grouping visual and functional differences. Since oligophrenia was recognized as a non-progredient condition, it was regarded as insufficient recovery from damage rather than as a progressive illness. The non-progressivity of oligophrenia, when it was considered to be the result of isolated exogenous causes, made oligophrenic development an endless process of recovery from a traumatic event.

The goal of achieving normal development over time expanded the possibilities for the correction and defined the boundaries for hopeless children. But in the absence of restrictions on biologically based explanations from the 1960s, oligophrenia could no longer be discussed in terms of mainly external causes. It became a complex condition in which genetic origins played a meaningful and later crucial part. One thing remained unchanged. Research methodology and expert descriptions were dominated by physiological, psychiatric, genetic, and other perspectives that made oligophrenia a subject of the natural sciences rather than the humanities.

The focus on deficits was strong and, in the 1960s and 1970s, research on oligophrenia became much more widespread, rich, and advanced. Although its key quality was the underdevelopment of abstract thinking, the aesthetic dimension of physiological defects as a visualisation of abnormality also defined the boundaries for the applicability of the term. Physiological deviations in psychophysiology served as lasting markers of multiple divergence from its course. The advancement of medical examinations together with statistics created a complex image of biologically rooted underdevelopment, understood as unfinished evolution.

References

Barashnev, Y. I., & Kazantseva, L. Z. (1975). Umstvennaya nedostatochnost' u detey, prozhivayushchikh v severnom izolyate. In A. A. Portnov (Ed.), *Aktual'nyye problemy oligofrenii* (pp. 150–162). Moskva.

Barashnev, Yu. I., & Yu. E. Vel'tishev. (1975). *Teoreticheskaya meditsina i pediatricheskaya praktika (Klinicheskaya genetika)*. Moskva.

Barashnev, Yu. I. et al. (1983). Nasledstvenno obuslovlennaya nedifferentsirovannaya oligofreniya u detey. In M. G. Blyumina et al. (Eds.), *Nasledstvennyye formy oligofrenii i drugikh anomaliy razvitiya u detey* (pp. 75–83). Moskva.

Bash, K. W., & Bash-Liechti, J. (1987). *Developing psychiatry. Epidemiological and social studies in Iran 1963–1976*. Springer.

Belkaniya, G. S., et al. (2017). *Lorin-Epshtejn Mojsej Yulijovich (1888–1945): zb. biografichnih, bibliografichnih ta naukovih materialiv*. Vinnicya.

Bell, M. (2010). Carl Gustav Carus and the science of the unconscious. In A. Nicholls, M. Liebscher (Eds.), *Thinking the unconscious. Nineteenth-century German thought*. Cambridge University Press.

Bezrogov, V. G., & Tendryakova, M. V. (2017). Chto nemtsu dialog, to russkomu uchitel': ot «Yestestvennoy istorii» G.K. Raffa k «Detskomu miru» K.D. Ushinskogo. In *Psikhologo-pedagogicheskiy poisk, 3*(43), 58–70.

Blumenbach, J. F. (1792). *An essay on generation*. Edinburgh.

Blumenbach, J. F. (1795). *Elements of physiology* (Vol. 1). Philadelphia.

Blumenbach, J. F. (1828). *The elements of physiology*. London.

Blyumenbakh, I. F. (1796). *Fiziologiya ili Nauka o yestestve chelovecheskom*. Moskva.

Blyumenbakh, I. F. (1797). *Rukovodstvo k yestestvennoy istorii*. Sankt Peterburg.

Blyumina, M. G. (1986). *Geneticheskoe izuchenie anomaliy razvitiya u deteiy*. Moskva.

Brailovskaya, K. (n.d.). Culturing the primitive mind: echoes of race in Soviet education.

Dupont, J.-C. (2007). Pre-Kantian revival of Epigenesis: Caspar Friedrich Wolff's De formatione intestinorum (1768–69). In P. Huneman (Ed.), *Understanding purpose: Kant and the philosophy of biology* (pp. 37–50). University of Rochester Press.

Efroimson, V. P., & Blyumina, M. G. (1978). *Genetika oligofreniy, psikhozov, epilepsiy*. Moskva.

Gambarotto, A. (2017). Lorenz Oken (1779–1851): Naturphilosophie and the reform of natural history. In *The British Journal for the History of Science, 50*(2), 329–340.

Gangus, A. (1982). *Riskovannoye priklyucheniye razuma*. Moskva.

Gaysinovich A. Ye. (1961). *K. F. Vol'f i ucheniye o razvitii organizmov*. Moskva.

Ginzburg, V. V. (1963). *Elementy antropologii dlya medikov*. Leningrad.

Goryaninov, P. (1837). *Zoologiya: osnovannaya na zoonomii i primenennaya k obshchey pol'ze, osobenno k meditsine, s prisoyedineniyem obshchey Organikologii i kratkoy Antropologii.* Sankt-Peterburg.

Grene, M. & David D. (2004). *The philosophy of biology: an episodic history.* Cambridge University Press.

Helbig, D., & Nassar, D. (2016). The metaphor of epigenesis: Kant, Blumenbach and Herder. In *Studies in history and philosophy of science Part A, 58*, 98–107.

Huneman, P. (2007). Reflexive judgment and Wolffian embryology: Kant's shift between the first and the third critiques. In P. Huneman (Ed.), *Understanding purpose: Kant and the philosophy of biology* (pp. 75–100). University of Rochester Press.

Jeannerod, M. (2006). The origin of voluntary action. History of a physiological concept. *C.R.Biologies, 329*, 354–362.

Kaasch, M., & Kaasch, J. (2003). Za i protiv Darvina—darvinisty i ikh protivniki sredi chlenov nemetskikh akademiy yestestvo ispytateley Leopol'dina. In G. S. Levit et al. (Eds.), *V teni darvinizma. Al'ternativnyye teorii evolyutsii v XX veke* (pp. 49–70). Sankt-Peterburg.

Karus, K. G. (1844). *Osnovaniya kranioskopii.* Sankt-Peterburg.

Kee, T. (2019). Blumenbach's race science in the light of Christian supersessionism. In N. Rupke & G. Lauer (Eds.), *Johann Friedrich Blumenbach race and natural history, 1750–1850* (pp. 123–140). Routledge.

Khit', G. L. (1983). *Dermatoglifika narodov SSSR.* Moskva.

Kirilov, N. (1845). *Karmannyy slovar' inostrannykh slov, voshedshikh v sostav russkogo yazyka, izdavayemyy N. Kirilovym.* S. Peterburg.

Kolmogorova, V. V., & Buykov, V. A. (2013). *Umstvennaya otstalost' (oligofreniya).* Chelyabinsk.

Krafft-Ebing. (1881–1882). Uchebnik psikhiatrii, sostavlennyy na osnovanii klinicheskikh nablyudeniy dlya prakticheskikh vrachey i studentov doktorom Kraft-Ebingom, T. 1–3. Sankt-Peterburg.

Lu-Adler, H. (2025). Not those who "all speak with pictures": Kant on linguistic abilities and human progress. In L. Filieri & K. Pollok (Eds.), *Kant on language.* Cambridge University Press.

Lubovskiy, V. I. (1955). *Nekotoryye osobennosti sovmestnoy raboty dvukh signal'nykh sistem v formirovanii dvigatel'nykh reaktsiy u detey-oligofrenov.* Moskva.

Luriya, A. R. (1958). Rol' rechi v regulyatsii normal'nogo i anomal'nogo povedeniya. In A. R. Luriya (Ed.), *Problemy vysshey nervnoy deyatel'nosti normal'nogo i anomal'nogo rebenka* (pp. 5–46). Moskva.

Luriya, A. R. (1963). *The mentally retarded child: Essays based on the study of the peculiarities of the higher nervous functioning of child-oligophrenics.* Macmillan.

Mates, P. (1927). *Konstitutsional'nyye tipy zhenshchin, v chastnosti interseksual'nyy tip.* Kiyev.

Mirskiy, M. B., & Blokhina, N. N. (2004). Nauchnyy vklad doktora Gaaza. In *Problemy sotsial'noy gigiyeny i istoriya meditsiny, 1*, 49–53.

Nicoglou, A. (2017). The timing of development. In C. Bouton & P. Huneman (Eds.), *Time of nature and the nature of time: Philosophical perspectives of time in natural sciences* (pp. 359–390). Springer.

Pevzner, M. S. (1959). *Deti-oligofreny (Izucheniye detey-oligofrenov v protsesse ikh vospitaniya i obucheniya).* Moskva.

Pevzner, M. S. (1961). *Oligophrenia: Mental deficiency in children.* Springer.

Pevzner, M. S. (1966). *Deti s otkloneniyami v razvitii (otgranicheniye oligofrenii ot skhodnykh sostoyaniy).* Moskva.

Pevzner, M. S. (1972). *Kliniko-geneticheskiye issledovaniya oligofrenii.* Moskva.

Pevzner, M. S., & Lebedinskaya, K. S. (1979). *Uchashchiyesya vspomogatel'noy shkoly.* Moskva.

Portnov, A. A. (1975). *Aktual'nyye problemy oligofrenii.* Moskva.

Pravdina-Vinarskaya, Ye. N. (1957). *Nevrologicheskaya kharakteristika sindroma oligofrenii.* Moskva.

Raff, G. C. (1778). *Naturgeschichte für kinder: mit elf kupffer-tafeln.* Göttingen.

Raff, G. K. (1796). *Yestestvennaya istoriya dlya maloletnikh detey: S 11 reznymi tablitsami*. Sankt-Peterburg.

Raykov, B. Ye. (1951). *Russkiye biologi-evolyutsionisty do Darvina. T.2*. Moskva.

Reed, E. W., & Reed, S. C. (1965). *Mental retardation: A family study*. Saunders.

Richards, R. J. (2000). Kant and Blumenbach on the Bildungstrieb: A historical misunderstanding. In *Studies in history and philosophy of science Part C: Studies in history and philosophy of biological and biomedical sciences, 31*(1), 11–32.

Richards, R. J. (2019). The beautiful skulls of Schiller and the Georgian girl. Quantitative and aesthetic scaling of the races, 1770–1850. In N. Rupke & G. Lauer (Eds.), *Johann Friedrich Blumenbach race and natural history, 1750–1850* (pp. 142–175). Routledge.

Richards, R. J. (2020). Romantic biology: Carl Gustav Carus at the edge of the modern. In E. Millán (Ed.), *The Palgrave handbook of German Romantic philosophy*. Palgrave Macmillan.

Richardson, A. (2001). *British romanticism and the science of the mind*. Cambridge University Press.

Russkih, V. V. (1969). *Oligofrenii i smezhnyye formy*. Moskva.

Severtsov, A. N. (1939). *Morfologicheskiye zakonomernosti evolyutsii*. Moskva-Leningrad.

Shmal'gauzen, I. I. (1982). *Organizm kak tseloye v individual'nom i istoricheskom razvitii*. Moskva.

Sjogren, T., & Larsson, T. (1970). A clinical and genetic population study of oligophrenia in a north Swedish isolate. A follow-up investigation thirty years later. *Acta Psychiatrica Scandinavica, 46*(s219), 198–208.

Sloan, P. R. (1979). Buffon, German biology, and the historical interpretation of biological species. In *British journal for the history of science, 12*(41), 109–153.

Steigerwald, J. (2019). *Experimenting at the boundaries of life. Organic vitality in Germany around 1800*. University of Pittsburgh Press.

Sukhareva, G. Ye. (1961). *Lektsii po oligofrenii 1*. Moskva.

Sukhareva, G. Ye. (1962). *Oligofreniya. Uchebnoye posobiye №1.*

Sukhareva, G. Ye. (1965). *Klinika oligofrenii (Klinicheskiye lektsii po psikhiatrii detskogo vozrasta)*. Moskva.

Troshin, G. Y. (1915). *Antropologicheskiye osnovy vospitaniya. Sravnitel'naya psikhologiya normal'nykh i nenormal'nykh detey, T.1*. Petrograd.

Ushinskiy, K. D. (1948). *Sobraniye sochineniy. T.2*. Moskva, Leningrad.

Vol'f, K. F. (1950). *Teoriya zarozhdeniya*. Moskva.

Vol'f, K. F. (1973). *Predmety razmyshleniy v svyazi s teoriyey urodov*. Leningrad.

Wallen, M. (2004). *City of health, fields of disease. Revolutions in the poetry, medicine, and philosophy of romanticism*. Routledge.

Yaroshevskiy, M. G. (1995). Uolter Kennon, Ivan Pavlov i problema emotsional'nogo povedeniya. *Voprosy psihologii, 6*, 55–65.

Yavkin, V. M. (1974). *Semeynyye formy oligofrenii*. Tashkent.

Zabugin, F. D. (1937). *Deti-oligofreny (imbetsily i idioty)*. Moskva.

Zammito, J. H. (2017). *The gestation of German biology: Philosophy and physiology from Stahl to Schelling*. University of Chicago Press.

Zankov, L. V. (1939). *Psikhologiya umstvenno-otstalogo rebenka*. Moskva.

Zharov, L. V., & Chernositova, T. L. (1989). Predisloviye. In F. P. Gaaz, *Moye puteshestviye na Aleksandrovskiye vody*. Stavropol'.

Chapter 8
Conclusion: Blumenbach's Legacy and the Continuing Reconsideration of Race in Science

Victoria Shmidt and Simone De Angelis

Abstract The final chapter discusses the challenges in historicizing scientific racism and the battle against it. The intellectual biographies of scholars who, like Blumenbach, were uncompromising fighters against the main political embodiment of racism in their time, bring about the question of the driving forces behind the mutually opposed political affiliations of scientists. The chapter brings forward a conceptualization racism and anti-racism in terms of immanent critique as a promising lens for addressing the task of historicizing race in science.

Keywords Immanent critique · Race in science · Racial intermixture · Anti-racism

Blumenbach is one of the first thinkers whose legacy opened up space for a mutually controversial historicization of the scientist's impact on scientific racism and the fight against it (Rupke & Lauer, 2018). A recent trend in the history of science has been to reexamine the legacy of scholars who were previously considered committed anti-racists, to identify their influence on the development of a racist worldview and the legitimization of racist practices. Franz Boas, who stood in direct opposition to openly racist anthropologists such as Nathaniel Southgate Shaler (Baker, 1998) and established the tradition of opposing a cultural anthropological perspective to physical anthropology, has recently become the target of such critical historicization (Pollock & Bernbeck, 2023). The wide range of criticism concerning Boas's legacy varied from rigorous adherence to his arguments and their scientific reliability (Allen, 1989) to revising the complex and internally conflicted positions of Boas regarding indigenous and Black scholars (Blackhawk & Wilner, 2018).

Comparable work can be recognized regarding many other scholars and thinkers, such as Ernest Haeckel (Lewit and Hossfeld Levit & Hossfeld, 2020; Richards, 2019),

V. Shmidt (✉) · S. De Angelis
Centre for the History of Science, University of Graz, Graz, Austria
e-mail: victoria.shmidt@uni-graz.at

S. De Angelis
e-mail: simone.de-angelis@uni-graz.at

V. Shmidt and S. De Angelis (eds.), *Blumenbach's Racial Classification*, History, Philosophy and Theory of the Life Sciences 39,
https://doi.org/10.1007/978-3-032-15950-2_8

Margareth Mead (Freeman, 1998), or Theodosius Dobzhansky (Gannett, 2013). The coexistence of historical evidence in favor of classifying the same scientist as both a racist and an anti-racist has become a source of frustration, clearly expressed in the questions of Martin Ericsson (2019, 156–157), biographer of the Swedish leftist geneticist Gunnar Dahlberg: "If so, is it possible that Dahlberg simultaneously battled racism, while in another way being partly responsible for building the foundations of that very same phenomenon?"

Exploring race in science often posits thinking framed *by* racism (and framing racially informed practices) against scientific arguments operating *against* racism (Gil-Riaño, 2023). Opposing race-informed and anti-racist arguments has reduced the history of race in science either to a transcendental timeline that treats the idea of race and racism as being "always" there or to reductive contextualism on the verge of relativism. The transcendental vision performs as categorical imperative, which relegates to on the margins the communicative nature of operationalizing race in science and the related task of historicizing it through mapping different interest groups in different periods in different contexts who produced pro-racist and anti-racist arguments. Relativism remains within the framework of disparate historical anecdotes, and even the comparison of which does not bring us closer to a dialectical understanding of the process of developing scientific arguments regarding race. Neither of these extremes can systematically contribute to the study of the role of scientific progress in explaining human diversity and racial politics.

Guided by moral criticism, the question of why Blumenbach embraced an abolitionist position, while other members of the Göttingen school continued to recognize slavery (Van Gorkom, 2022), often leads to a questionable and univocal interpretation of racial intermixture as an argument against racism. Apparently, Blumenbach's argument, based on the assumption of human variation as transient, attacked the theoretical foundations of slavery, the central institution of racism in Blumenbach's time. The intersectionality of race and class, which shaped multiple policies aimed at applying selection among and within minorities, typical of the post-slavery imperial period, should be seen as one of the crystallized, pro-racist elements of Blumenbach's legacy. Along with racial classification and skull collection, which already brought Blumenbach's racism into the focus of debates, his transitory view on races laid down the track for approximating the Caucasian race as the most desirable for the rest of humanity.

Blumenbach's straightforward influence on the next generations of thinkers was coined in the debates regarding either the positive or the negative role of racial intermixture in the destiny of humanity. The division into adherents and critics of racial intermixture persisted as a main arena for producing arguments for and against surveillance over the reproduction of those seen as the genetic load for nations, and even for humanity. Even though the legitimization of extermination in the name of racial purity is a well-known part of the historiography of race in science, the strict surveillance of "isolated" groups nevertheless requires a more systematic historiography.

Remarkably, the boundary between supporters and opponents of racial mixing was and remains very blurred. Even those who came out as the most active champions of racial purity, like the ideologist of Racial Hygiene, Eugene Fisher, recognized mixed European races as the result of a positive development: Fisher himself considered himself to be of the "newly discovered" Dinaric race (Hoffman, 1937, 163). Conversely, Eastern European anthropologists like Božo Škerlj, who proclaimed racial mixing as the only possible trajectory for human progress (Škerlj, 1934a), could not accept mixing with Jews, especially those who were seen as genetically and culturally isolated from progressively developing humanity: "I am for free competition between races, although [...] I could not change my opinion about the Jews, [...] and I would not allow these people free competition with our European races." (Škerlj, 1934b) Not occasionally, Škerlj was one of the popularizers of the Dinaric "race" and even a student of Fischer during his stay at the *Kaiser-Wilhelm-Institut für Anthropologie, menschliche Erblehre und Eugenik.*

Such a relational understanding of views on racial intermixture became even more complex when the population groups seen as racially pure and isolated at the end of the 1950s, started to be considered valuable for discovering the mystery of human origins. Accompanied by an audible motif of extinction, studies of isolates resulted in the complex stigma of these populations being known as "non-adaptable." The research on "isolated groups" determined the agenda of several international research networks, including the International Biological Program (Weiner, 1964) and the *Rassengeschichte der Menschheit* (*The Racial History of Mankind*) initiated by Karl Saller in the 1960s and finished by Ilse Schwidetzky in the 1980s (Saller & Schwidetzky, 1968). These projects initiated the application of systematic analogies between ethnic groups such as European Roma, Innuits, and various tribes in Latin America and Africa, which directly shaped a new wave of geneticization of these groups, seen as less adaptable due their "complex isolation." (Shmidt & Jaworsky, 2020).

Obsession with the search for a scientifically proven and genetically informed explanation of adaptability brought biologists face to face with a further dichotomy, namely, between classical and balance explanations for genetic variation. Even the contest between different approaches to the role of natural selection in human variety was established earlier, after 1945, attributing a decisive role to either undesirable mutations or desirable diversity, responded to the new, post-genocide challenges of genetics. The awarding of the Nobel Prize in Biology in 1946 to Hermann J. Muller for the discovery of mutations due to radiation became one of the drivers of the escalation of the rift between him and the geneticists led by Theodosius Dobzhansky, who was confident about the positive role of mutations.

Muller's conceptualization in terms of genetic load and its various manifestations operated in favor of legitimizing negative eugenics (Calabrese & Shamoun, 2025). In his Nobel lecture, Muller (1946) directly linked the high probability of more rapid genetic degeneration with the higher frequency of genetic mutations, opening the door for justifying surveillance over such risks. In his turn, Dobzhansky challenged Muller's idea of overdominance as a strategy for improving humanity and insisted that variety would be a core prerequisite for the survival of humanity (Crow, 1987). The Muller-Dobzhansky controversy, which nurtured an adaptationist agenda, was confronted by Richard Lewontin, who revised both classical and balance explanations of genetic variation, with the assistance of new discoveries in molecular biology (Winther, 2021, 10–11).

The pitfalls of adaptationism, including its high potential for producing race-informed arguments, shaped one of the most recent anti-racist agendas in human biology, developed by Stephen Jay Gould, Richard C. Lewontin, and Richard Levins. One of the starting points for the destruction of adaptationism was the mixing of different notions of adaptation, which led to confusion over the social and biological (genetic) signifiers of adaptation (Lewontin, 1979). The proponents of adaptability emphasized the central, albeit non-monopolistic, role of adaptation as crucial for human evolution.

In their turn, Gould and Lewontin focused on the historical significance of non-adaptation, or rather on which profile of reproduction could be defined as inconsistent with the expectations of adaptationism (Gould & Lewontin, 1979). Anti-adaptationist programs included multiple notices against the predominance of the biological deterministic view on human diversity (Shen & Feldman, 2022), and very limited options for validating historical adaptive hypotheses (Nielsen, 2009). The debates between adaptationists and anti-adaptationists, along with the development of anti-adaptationist programs, continued to align with further progress in the research on human variation and human origins (Godfrey-Smith, 1999; Lewens, 2009).

The heated debates between the proponents and opponents of adaptation echo the tension between Blumenbach and Meiners, the proponents and opponents of racial mixing such as Prichard and Nott, those who trusted in the power of assimilation like Boas, and those who rejected such ideas regarding the many migrants who arrived in the United States from Central Europe. However, the ongoing reproduction of the pattern of biologists using their knowledge to defend mutually exclusive views on the politics that form the core of racism in a given period is not the only direction in understanding the continuities and interruptions of the history of race in science. The destiny of Blumenbach's approach to human variation, along with praise for racial intermixture, fighting for preserving isolates, and improving human adaptability, calls for investigating the explanatory importance of the social structures that construct race in the production of knowledge. Recognizing race in science as the product of a plurality of such structures involves the deconstruction of dichotomies between "bad" and "good" science.

Through revising the controversy between Muller and Dobzhansky, as reflecting their commitment to explain behavioral differences by genetic difference, Diane Paul (1987) provides more general insights regarding the task of exploring the driving forces that link scientific and social values. For Paul, dedication to the ideal of bettering humanity operated as a filter for scientific tasks. The question of what gives rise to the tendency to deconstruct such filters leads to the mechanisms of immanent critique within knowledge production that aims to explain human diversity.

Phillip Sloan (2006) reviews Kant's contribution to the development of better methodologies in biology and draws attention to the role of sensitivity to the distance between the descriptive category of variety and the explanatory concept of race, which aims at setting boundaries. In his detailed revision of Kant's input, Sloan reconstructs the driving forces that directed Kant to practicing immanent critique regarding the methodologies to produce biological knowledge, which he encouraged but also recognized as problematic. Bringing in the focus of his historicization of the social ontology of Kantian critique Sloan discusses important facts previously relegated to the margins of Kant's intellectual biography, such as the development of Buffon's distinction of "abstract" and "physical" truth. Moreover, in Sloan's reading of the debates between Kant and Forster, regarding the category of race, widely examined hy historians of race (Strack, 1996) are not just one more clearly unsuccessful attempt to emancipate scientific argument from moral judgement. Through following the formation of a Kantian critique of Forster's and Herder's approaches to race through spatialized time, Sloan convincingly demonstrates the role of Kant in the formation of not only individual methodologies but also the positioning of relations between them. Sloan's historicization of Kant successfully confronts the dichotomy of contextualism vs. unification with regard to race (Ludwig, 2019). Furthermore, it shows that moving away from what David Ludwig defines as "simple contextualism" in historicizing race toward producing an entangled history of racial ontologies is a promising strategy.

Any discussion of the options for exploring the entangled history of race in science calls for geopolitical approximation, vital for connecting local, inter-country, transnational, and global contexts. Accepting this challenge requires explicating a historically informed vision of such geopolitical approximation. In this vein, the main purpose of historicizing race in science is to reconstruct and explore the complex interrelationship between the pros and cons of social practices associated with racism, such as slavery, racially informed surveillance, and genocide.

If there is a possibility to find the solution to move beyond the dichotomy between racism and universalism of its negation, it should be historically and geopolitically contextualized, not least because of the high probability of being adopted by proponents of racism. One of the main methodological challenges in examining racism and anti-racism is to "to move beyond both reductionism and simple non- or anti-reductionism." (Bhaskar & Danermark, 2006, 280). To be precise, racialized arguments have played an important role, and they continue to be one of the main driving forces in the ongoing reproduction of colonialism. Meanwhile, anti-racism itself stands out as a racially charged practice; Blumenbach's case clearly affirms this fact.

Further, approaching these tasks with the assistance of critical historicization inclines us to follow the path of complex historicization, which would solve global-level problems in the context of a critical rethinking of the history of certain geopolitical clusters. The entangled history of race in science contains hitherto unappreciated or even unknown transnational connections that can only be revealed through a continuous reconsideration of our approaches to mapping progress in science and its interplay with colonial power.

References

Allen, J. S. (1989). Franz Boas's physical anthropology: The critique of racial formalism revisited. *Current Anthropology, 30*(1), 79–84.

Baker, L. D (1998). "Columbia University's Franz Boas: He Led the Undoing of Scientific Racism." *The Journal of Blacks in Higher Education, 22*, 89–96.

Bhaskar, R., Danermark, B. (2006). Metatheory, interdisciplinarity and disability research: A critical realist perspective. *Scandinavian Journal of Disability Research, 8*(4), 278–297.

Blackhawk, N., & Wilner, I. L. (2018). *Indigenous visions: Rediscovering the world of Franz Boas.* Yale University Press.

Calabrese, E. J., & Shamoun, D. Y. (2025). The unraveling of a Nobel prize: How Hermann Muller was awarded the Nobel prize: A front for eugenics. *Journal of Occupational and Environmental Hygiene, 22*(3), 149–168.

Crow, J. F. (1987). Muller, Dobzhansky, and overdominance. *Journal of the History of Biology, 20*(3), 351–380.

Ericsson, M. (2019). "Anti-fascist race biology: Gunnar Dahlberg and the long farewell to the Nordic 'master race'." In K Braskén, N. Copsey, & J. A. Lundin (Eds.), *Anti-Fascism in the Nordic Countries: New Perspectives, Comparisons and Transnational Connections* pp. 143–159. Routledge.

Freeman, D. (1998) *The fateful hoaxing of Margaret Mead: A historical analysis of her Samoan research.* Westview Press.

Gannett, L. (2013). Theodosius Dobzhansky and the genetic race concept. *Studies in History and Philosophy of Science Part C: Studies in History and Philosophy of Biological and Biomedical Sciences, 44*(3), 250–261.

Gil-Riaño, S. (2023). *The remnants of race science: UNESCO and economic development in the global south.* Columbia University Press.

Godfrey-Smith, P. (1999). Adaptationism and the power of selection. *Biology and Philosophy, 14*, 181–194.

Van Gorkom, J. (2022). "Johann Gottlieb Stoll and Forster's Challenge to Kant." *Critical Philosophy of Race, 10*(2), 295–311.

Gould, S., & Lewontin, R. (1979). "The spandrels of San Marco and the panglossian paradigm: A critique of the adaptationist programme." *Conceptual Issues in Evolutionary Biology, 205*, 79.

Hoffman, M. (1937). *Heads and tales in many lands.* Scribner.

Levit, G. S., Hossfeld, U. (2020). Ernst Haeckel, Nikolai Miklucho-Maclay and the racial controversy over the Papuans. *Frontiers in Zoology,* 17, 1–20.

Lewens, T. (2009). Seven types of adaptationism. *Biology & Philosophy, 24*, 161–182.

Lewontin, R. (1979). Sociobiology as an adaptationist program. *Behavioral Science, 24*(1), 5–14.

Ludwig, D. (2019). How race travels: Relating local and global ontologies of race. *Philosophical Studies, 176*(10), 2729–2750.

Muller, H- J. (1946). "The production of mutations" Nobel Prize lecture, Stokholm.

Nielsen, R. (2009). Adaptionism—30 years after Gould and Lewontin. *Evolution, 63*(10), 2487–2490.

Paul, D. B. (1987). Our load of mutations revisited. *Journal of the History of Biology, 20*(3) 321–335.

Pollock, S., & Bernbeck, R. (2023). Franz Boas: Between anti-racism and reification. *EAZ–Ethnographisch-Archaeologische Zeitschrift, 57*(1). https://doi.org/10.54799/RGBN4865.

Richards, R. J. (2019). *The tragic sense of life: Ernst Haeckel and the struggle over evolutionary thought.* University of Chicago Press.

Rupke, N., & Lauer,G. (2018). "Introduction: A brief history of Blumenbach representation." In N. Rupke, & G. Lauer (Eds.), *Johann Friedrich Blumenbach* pp. 3–15. Routledge.

Saller, K., & Schwidetzky, I. (1968). *Rassengeschichte der Menschheit.* R. Oldenbourg.

Shen, H., & Feldman, M. W. (2022). "Diversity and its causes: Lewontin on racism, biological determinism and the adaptationist programme." *Philosophical Transactions of the Royal Society, 377*(1852)

Shmidt, V., & Jaworsky, B. N. (2020). *Historicizing Roma in Central Europe: Between critical whiteness and epistemic injustice.* Routledge.

Škerlj, B. (1934a, December 02). "Die Rassen Zentraleuropas: Ergebnisse der anthropologischen Forschung in Jugoslavien" [The races of Central Europe: Outcomes of anthropological research in Yugoslavia] Prager Presse.

Škerlj, B. (1934b, November 03). Letter to Jindřich Matiegka. Archive of the Group of Anthropology, Department of Biology, Biotechnical Faculty, University of Ljubljana, Fond of Božo Škerlj.

Sloan, P. R. (2006). Kant on the history of nature: The ambiguous heritage of the critical philosophy for natural history. *Studies in History and Philosophy of Science Part C: Studies in History and Philosophy of Biological and Biomedical Sciences, 37*(4), 627–648.

Strack, T. (1996). Philosophical anthropology on the eve of biological determinism: Immanuel Kant and Georg Forster on the moral qualities and biological characteristics of the human race. *Central European History, 29*(3), 285–308.

Weiner, J. S. (1964). "The biology of man in the international biological programme: The human adaptability project". *Current Anthropology 5*(3), 191–195.

Winther, R. G. (2021). Lewontin (1972). In *Remapping race in a global context* (9–47). Routledge.

Index

V. Shmidt and S. De Angelis (eds.), *Blumenbach's Racial Classification*, History, Philosophy and Theory of the Life Sciences 39,
https://doi.org/10.1007/978-3-032-15950-2

MIX
Papier aus verantwortungsvollen Quellen
Paper from responsible sources
FSC® C105338
FSC
www.fsc.org

If you have any concerns about our products,
you can contact us on
ProductSafety@springernature.com

In case Publisher is established outside the EU,
the EU authorized representative is:
Springer Nature Customer Service Center GmbH
Europaplatz 3, 69115 Heidelberg, Germany

Printed by Libri Plureos GmbH
in Hamburg, Germany